MEDICAL BIOINFORMATICS AND BIOCHEMISTRY

(DIABORMATICS)

BIOCHEMISTRY RESEARCH TRENDS

Additional books and e-books in this series can be found on Nova's website under the Series tab.

BIOCHEMISTRY RESEARCH TRENDS

MEDICAL BIOINFORMATICS AND BIOCHEMISTRY

(DIABORMATICS)

RAJNEESH PRAJAPAT
AND
IJEN BHATTACHARYA

EDITORS

NOTICE TO THE READER

Library of Congress Cataloging-in-Publication Data

ISBN: 978-1-53614-952-4
Library of Congress Control Number: 2019931108

Published by Nova Science Publishers, Inc. † New York

CONTENTS

PREFACE

Diabetes is a group of complex metabolic disorders principally characterized by insulin resistance, pancreatic beta cell dysfunction, and associated hyperglycemia. This disease is increasing to epidemic proportions in many countries throughout the world. Diabetes is also associated with accelerated atherosclerotic macro vascular disease affecting arteries that supply the heart, brain and lower extremities. As a result, patients with diabetes have higher risk of myocardial infarction, stroke and limb amputation. Large prospective clinical studies show a strong relationship between glycaemia and diabetic microvascular complications in both type 1 and type 2 diabetes. Hyperglycaemia and insulin resistance both seem to have important roles in the pathogenesis of macrovascular complications.

Molecular Docking continues to be a great promise in the field of computer based drug design. Several simulation models have been proposed to study the physiology and pathophysiology of diabetes. Biological databases and Atlas plays an important role in getting up-to-date global report on diabetes. Like so many other areas of medicine, Bio informatics has had a profound impact on diabetes research. The DPP-4 inhibitors are a newer class of drugs that lowers glucose by blocking an enzyme, thereby prolonging incretion effect *in vivo*.

This book is based on the recent development in the research dynamics of medical bioinformatics, biochemistry and the progress in these fields. It also provides current reference material for students entering in the field of medical biochemistry and bioinformatics academics and research as well as scientists already familiar with the area. The development in genomic sequencing and *in silico* biology has provided the data needed to accomplish comparisons of derived protein sequences, the results of which may be used to formulate and test hypotheses about biochemical function. The book is a useful source of knowledge for MBBS, BSc, MSc/MD/MS and PhD level students looking for an accessible introduction of the subject.

In: Medical Bioinformatics and Biochemistry   ISBN: 978-1-53614-952-4
Editors: Rajneesh Prajapat et al.   © 2019 Nova Science Publishers, Inc.

Chapter 1

DIABETES AND DIABORMATICS: AN INTRODUCTION

Rajneesh Parapet*[*]*, Ijen Bhattacharya
and Uday Kumar Gupta
Department of Medical Biochemistry,
Faculty of Medical Sciences, Rama Medical College
and Hospital (NCR), Rama University, Kanpur, UP, India

ABSTRACT

Bioinformatics study includes analysis of biological sequence data for the identification of structure and function of macromolecules. The structural biochemistry involves bioinformatics that deals with sequence alignments are obtained and eventually how the analysis of the sequences can help generate phylogenetic trees. Biochemistry involves the study of the chemical processes that occur in living organisms with the aim of understanding the nature of life in molecular terms. Diabormatics includes the bioinformatics aspect of diabetes genetics, drug development and precision medicine.

---

[*] Corresponding Author's Email: prajapat.rajneesh@gmail.com.

Diabetes mellitus (DM) is a complex metabolic disorder that characterized by high level of blood glucose due to insufficiency or ineffective secretion of pancreatic insulin hormone. The DM classified into two categories includes, type-1 diabetes and type-2 diabetes. Type-1 diabetes is an autoimmune disease resulting in destruction of pancreatic cells leading to severe lack of insulin. Whereas type-2 diabetes develops due to inefficient insulin utilization referred as insulin resistance or insufficient quantity of insulin production.

Insulin signaling pathway is the key pathway involved in regulating blood glucose level. Diabetes treatment aims at controlling blood glucose level. There are various kinds of chemical drug and herbal/natural products being used to effectively control blood glucose level.

Keywords: diabetes mellitus, insulin, diabormatics

INTRODUCTION

Bioinformatics or computational biology is an inter-disciplinary branch of study, that deals with application of mathematical, statistics, biology and informational techniques principles to process the biological data to find out the structure and functions of macro-molecules (e.g., DNA, RNA and protein) and its applications that cover drug discovery, genome analysis and biological control of gene expression and protein interactions, etc. [1].

Biochemistry involves the study of the chemical processes that occur in living organisms with the aim of understanding the nature of life in molecular terms. Biochemical studies use analytical techniques for understand about natural phenomenon and relationships between biological molecules, such as nucleic acids (DNA, RNA) proteins and cellular function. The biochemical analysis informs about the complexity of biological systems provides an excellent source of extremely large data sets for bioinformatics analysis [2].

Biochemical and bioinformatics studies have enlightened many aspects of health and disease, and the study of various aspects of health and disease has opened new areas of research. Biochemistry makes significant contributions to the fields of cell biology, physiology, immunology, microbiology, pharmacology, and toxicology, as well as the fields of inflammation, cell injury, and cancer. These close relationships emphasize that life, as we know it, depends on biochemical reactions and processes.

"Diabormatics is an interdisciplinary area of study that includes the bioinformatics analysis of diabetes genetics, drug development and precision medicine."

1.1. DIABETES: AN INTRODUCTION

Diabetes has emerged as a major healthcare problem worldwide. Diabetes mellitus (DM) is a metabolic disorder caused due to insufficient or ineffective insulin and characterized by high blood glucose [3] The Greek physician Arteus (ca A.D. 200) was call insulin deficiency disease as diabetes. The different types of DM are caused by a complex interaction of genetics and environmental factors. In 2007, there was an estimated 40 million persons suffering from diabetes in India and this number possibly rises to almost 70 million people by 2025. By 2030, countries with the largest number of diabetic people will be India, USA and China [4].

There are two types of diabetes mellitus namely, type-1 diabetes (T1DM) [10% of total diabetic population] and type-2 diabetes (T2DM) [90% of total diabetic population] [5, 7]. Type-1 diabetes is an autoimmune disorder that results in destruction of β-cells in the islets of Langerhans (Pancreas) resulting in a decrease of insulin hormone production [6]. The diagnosis and treatment of T1DM involve monitoring of blood glucose level and administration of insulin, regular exercise and diet management [8].

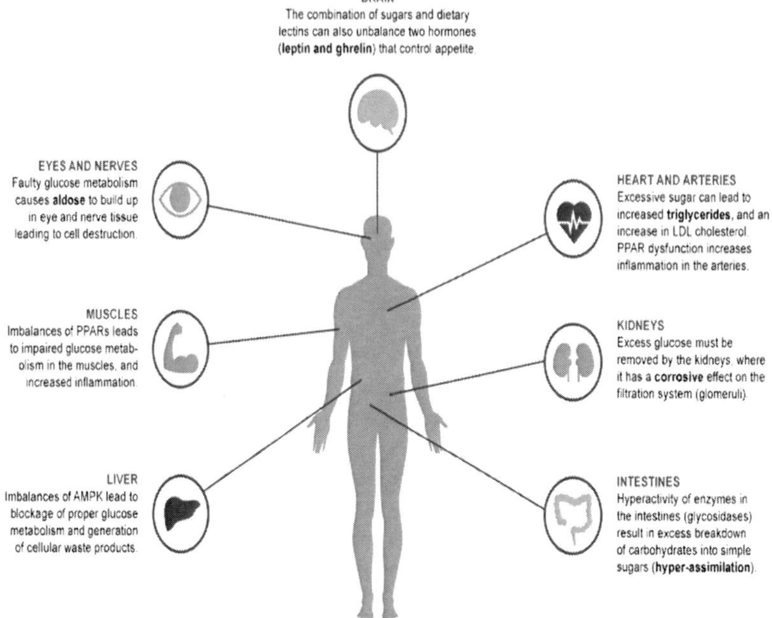

Figure 1.1. Consequences of sugar imbalance.

T2DM is also called as non-insulin dependent diabetes mellitus (NIDDM) [9, 10]. T2DM is due to a combination of defective secretion of insulin from pancreatic β-cells and impairment of insulin- mediated glucose disposal, which is called insulin resistance, or the pancreas do not produce sufficient quantity of insulin [9, 10]. The diagnosis and treatment of T2DM involve maintenance of blood glucose level in control by healthy diet, regular exercise and lifestyle changes as required.

1.2. ROLE OF INSULIN AND GLUCAGON IN REGULATING BLOOD GLUCOSE LEVEL

The two pancreatic hormones - insulin and glucagon regulate the blood glucose levels. The islets of Langerhans are the polygonal bunch of pancreatic cells which produce insulin and glucagons [11, 12]. In insulin structure, two polypeptide chains are held together by cross linkages of two disulfide bonds. The acidic chain A contains 21 residues and the peptide chain B having 30 residues.

Increased blood glucose levels (hyperglycemia) trigger beta cells of islets of Langerhans to secrete insulin, which enhances glucose uptake by the cells (Figure 1.3). Inside the cell, the glucose is either converted into energy and used by the cell or is converted to glycogen and stored mainly in the liver and skeletal muscles or is used for the production of fats. Low blood glucose level or hypoglycemic condition triggers alpha cells of islets of Langerhans to produce glucagon [11, 12]. Glucagon stimulates the breakdown of glycogen to glucose, which is then released into the bloodstream [11, 12] (Figure 1.2).

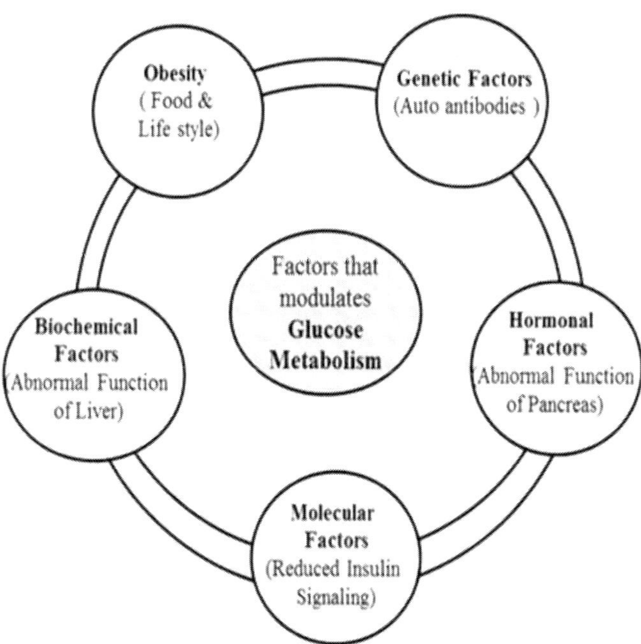

Figure 1.2. Summary of the factors modulating Glucose metabolism.

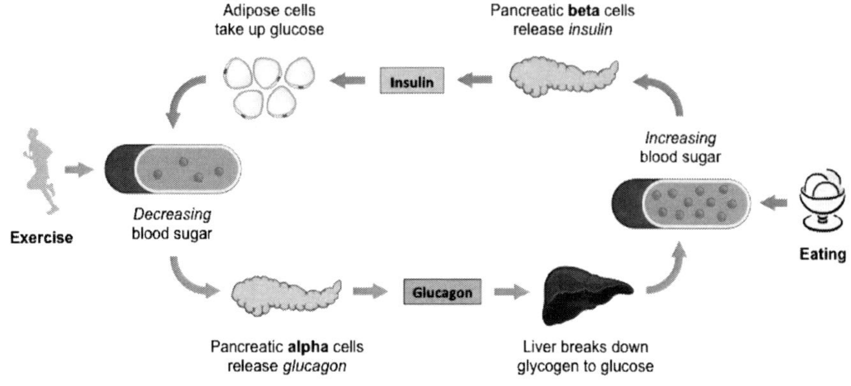

Figure 1.3. Role of insulin and glucagon in regulating blood glucose levels.

1.3. MEDICATIONS FOR DIABETES

Traditional Indian herbal drugs and plants used in the treatment of diabetes. Following medicinal plants has antidiabetic property and related beneficial effects and of herbal drugs used in treatment of diabetes. This includes *Eugenia jambolana, Momordica charantia Ocimum sanctum, Allium sativum, Pterocarpus marsupium, Withania somnifera, Tinospora cordifolia*, and *Trigonella foenum* [13]. The prescribed drugs for diabetes mellitus act on controlling blood glucose levels either by enhancing insulin secretion or regulating glucose metabolism. Administration of insulin is the most prescribed therapeutic method for patients with type-1 diabetes [14].

Drugs like GLP-1 analogues and DPP-4 inhibitors act by stimulating insulin production and retarding glucagons production (Figure 1.4). Some of the side effects which have been reported for the chemical antidiabetic drugs are low blood glucose level (Sulphonylureas), stomach discomfort (Alpha-glucosidase Inhibitors, DPP-4 Inhibitors) and respiratory infection (DPP-4 Inhibitors).

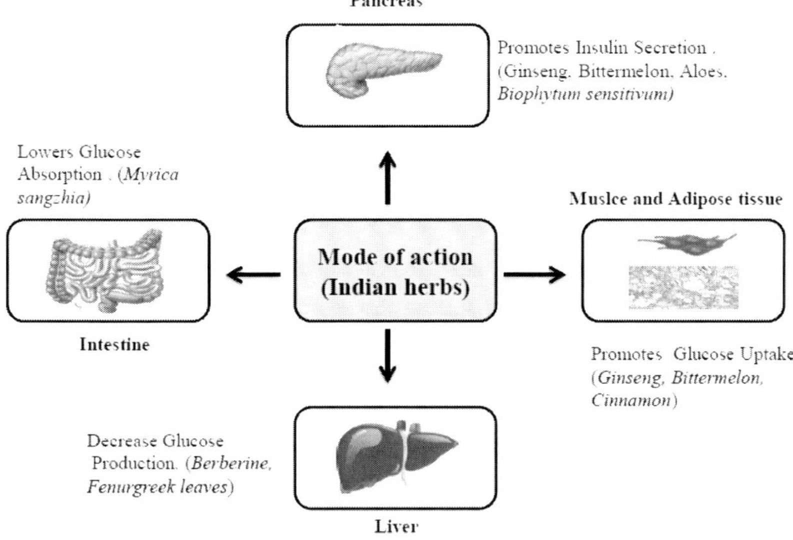

Figure 1.4. Mode of action of some of the anti-diabetic medicinal plants.

Traditional Medicines derived from medicinal plants are used by Indian population. Use of traditional therapeutic methods like use of medicinal plants is the most widely used approach for primary health care across the globe. As diabetes is a global disease which requires lifelong monitoring and medication the anti-diabetic potential of several herbal plants has been experimentally tested and proven. Many of these plant extracts are being effectively used either alone or in combination with conventional therapeutic methods for efficient treatment and control of diabetes. The high cost of chemical drugs, lifelong dependency, along with its side effects has made the alternative therapeutic medicines even more popular. Phytochemicals obtained from the medicinal plants aid in developing formulations for synthetic or semi- synthetic drugs.

The herbal medications control blood glucose levels by the one of the following mechanisms: increase insulin secretion, increase insulin sensitivity, inhibit gluconeogenesis and glycogenolysis in liver, stimulate glycogenesis, enhance uptake of glucose in adipose and skeletal muscles tissues, inhibiting absorption of glucose in the intestine, improve glucose metabolism, reduce lipid per-oxidation, reduce oxidative stress (Figure 1.4) [15, 16]. The major problem with this herbal formulation is that the active ingredients are not well-defined. The active component and their molecular interaction help to standardize and analyze therapeutic efficacy of product.

CONCLUSION

Diabetes mellitus is a complex metabolic disorder resulting from either insulin dysfunction or insufficiency, and there is an ever-increasing rate for diabetes worldwide due to unhealthy diet, sedentary lifestyle and excessive body weight. There are various reported factors that alter insulin secretion thus resulting in etiology and progression of diabetes. Medicinal plants are being used for the treatment of diabetes. Managing blood glucose levels by medication, diet and exercise is the key for diabetes management and control. Biochemical and bioinformatics studies have enlightened many aspects of health and disease. Biochemistry makes significant contributions to the fields of cell biology, physiology, immunology, microbiology, pharmacology, and toxicology, as well as the fields of inflammation, cell injury, and cancer. Diabormatics is a new interdisciplinary area of study that includes the bioinformatics analysis of diabetes genetics, drug development and precision medicine. Understanding the disease and lifestyle changes along with incorporation of anti-diabetic plants as diet supplement will prevent, delaying the onset and effective management of diabetes. The herbal medicine in modern medical practices is lack of scientific and clinical data proving their efficacy and safety.

REFERENCES

[1] Luscombe NM, Greenbaum D, and Gerstein M. 2001. "What is bioinformatics? A proposed definition and overview of the field." *Methods Inf Med.* 40: 346-58.

[2] Wilson K, and Walker J. 2010. *Principles and Techniques of Biochemistry and Molecular Biology.* Cambridge University Press, New York.

[3] Diagnosis and classification of diabetes mellitus. *Diabetes Care.* 2009; 32 Suppl 1:S62-67.

[4] King H, Aubert RE, and Herman WH. 1998. "Global burden of diabetes, 1995-2025: prevalence, numerical estimates, and projections." *Diabetes Care* 21(9):1414-1431.

[5] Azar ST, Tamim H, Beyhum HN, Habbal MZ, and Almawi WY 1999. "Type I (insulin-dependent) diabetes is a Th1- and Th2-mediated autoimmune disease." *Clin Diagn Lab Immunol* 6 (3):306-310.

[6] Maahs DM, West NA, Lawrence JM, and Mayer-Davis EJ. 2010. "Epidemiology of type 1 diabetes." *Endocrinol Metab Clin North Am.* 39(3):481-497.

[7] Diagnosis and classification of diabetes mellitus. *Diabetes Care.* 2005; 28 Suppl 1:S37-42.

[8] Taylor R. 2012. "Insulin resistance and type 2 diabetes." *Diabetes* 61(4):778-779.

[9] Olokoba AB, Obateru OA, and Olokoba LB. 2012. "Type 2 diabetes mellitus: a review of current trends." *Oman Med Journal* 27(4):269-273.

[10] Taniguchi CM, Emanuelli B, and Kahn CR. 2006. "Critical nodes in signalling pathways: insights into insulin action." *Nat Rev Mol Cell Biol.* 7(2):85-96.

[11] Lefebvre PJ. 1995. "Glucagon and its family revisited." *Diabetes Care* 18(5):715-730.

[12] Kruger DF, Martin CL, and Sadler CE. 2006. "New insights into glucose regulation." *Diabetes Educ.* 32(2):221-228.

[13] Modak M, Dixit P, Londhe J, Ghaskadbi S, and Devasagayam PAT. 2007. Indian Herbs and Herbal Drugs Used for the Treatment of Diabetes. *Journal of Clinical Biochemistry and Nutrition* 40: 163–173. http://doi.org/10.3164/jcbn.40.163.

[14] Kraegen EW, Cooney GJ, Ye JM, Thompson AL, and Furler SM. 2001. "The role of lipids in the pathogenesis of muscle insulin resistance and beta cell failure in type II diabetes and obesity." *Exp Clin Endocrinol Diabetes* 109 Suppl 2:S189-201.

[15] Senn JJ, Klover PJ, Nowak IA, and Mooney RA. 2002. "Interleukin-6 induces cellular insulin resistance in hepatocytes." *Diabetes* 51(12):3391-3399.

[16] Lowell BB, and Shulman GI. 2005. "Mitochondrial dysfunction and type 2 diabetes." *Science* 307(5708):384-387.

In: Medical Bioinformatics and Biochemistry   ISBN: 978-1-53614-952-4
Editors: Rajneesh Prajapat et al.   © 2019 Nova Science Publishers, Inc.

Chapter 2

COMBINED EFFECT OF VITAMIN C AND E DOSE ON TYPE 2 DIABETES PATIENTS

Rajneesh Prajapat[1,], Ijen Bhattacharya[1]*
and Anupam Jakhalia[2]
[1]Department of Medical Biochemistry,
Faculty of Medical Sciences, Rama Medical College
and Hospital (NCR), Rama University, Kanpur, UP, India
[2]Department of Medical Biochemistry, Faculty of Medical Sciences,
Jawaharlal Nehru Medical College, Ajmer, Rajasthan, India

ABSTRACT

Diabetes is a metabolic disorder that causes vascular complications. As vitamin C and E is known for its beneficial effects on blood sugar, serum lipids and glycated haemoglobin (HbA1c). In the present study, we assess the combined effect of vitamin C and E on blood sugar (FBS), serum creatinine (SC), total cholesterol (TC), low and high density lipoprotein (LDL, HDL), and glycated haemoglobin (HbAIc) in type 2 diabetes mellitus patients. A total of 50 patients with type 2 diabetes referred to

---

[*] Corresponding Author's Email: prajapat.rajneesh@gmail.com.

Rama Hospital (NCR), India, were included in the study. They received 500 mg daily twice of both vitamin C and E for 4 months. Fasting blood sugar (FBS), serum creatinine (SC), total cholesterol (TC), low and high density lipoprotein (LDL, HDL), and HbAIc were measured before and after vitamin C and E consumption and the results were analyzed. A significant decrease in FBS, TC level and non-significant decrease in SC, LDL, and HbA1c level was seen in the patients supplemented with 500 mg of both vitamin C and E twice in a day for 4 months. Results indicate that daily consumption of 500 mg of vitamin C and E for 4 months may be beneficial for decreasing the FBS, TC, SC, LDL, and HbA1c and slight raise in HDL and calcium level in patients with type II diabetes and thus reducing the risk of complications.

Keywords: diabetes, HbA1c, LDL, HDL, Vitamin

INTRODUCTION

Pandit and Pandey [1]; Awasthi et al. [2] proposed that diabetes mellitus is a multi-factorial metabolic disorder and forthcoming epidemic all over the globe that caused due to ineffective secretion of insulin. Prajapat and Bhattacharya [3]; Zhaolan et al. [4] proposed that diabetic patient numbers will possibly rise up to 300 million by 2025 in India. Seyed Hosseini et al. [5] and Rahman Hassan et al. [6] also estimated that 346 million diabetic patients will be increase up to 439 million in 2030 worldwide.

In another study, Khabaz et al. [7] and Manzella et al. [8] explained that, vitamin E supplementation could improve glycemic control. Afkhami-Ardekani et al. [9] and Dakhale et al. [10] proposed that, vitamin C and glucose show structural similarities, and thus they are effective in prevention of non-enzymatic glycosylation of proteins. Chambial et al. [11] proposed that, vitamin C is an antioxidant that protects body from damage caused by free radical and also used as therapeutic agent for diseases and disorders. In the study, Simom [11] proposed that, vitamin C acts as regulator of catabolism of cholesterol and Ness et al. [13] proposed beneficial effects on lipids regulation.

In another study, Battisi et al. [14] and Manjunath et al. [15] proposed that, most of the patients with diabetes have lipid metabolism disorders; most prevalent forms are decreased high density lipoprotein (HDL) and increased triglyceride. Errikson and Kahvakka [16] proposed that, high doses of ascorbic acid (2 gm/day) improved blood glucose regulation and reduce serum cholesterol and triglyceride in type 2 diabetes patients. In another study, Sargeant et al. [17] proposed that, in year 2000 scientist reported an inverse relationship between mean plasma vitamin C and HbA1c levels. We undertook this study to evaluate the effects of vitamin C and E supplement on fasting blood sugar (FBS), serum creatinine (SC), total cholesterol (TC), low and high density lipoprotein (LDL, HDL) and HbAIc in patients with type 2 diabetes.

2.1. USE OF VITAMIN C AND E FOR DIABETES TREATMENT

The study was performed at the Rama Hospital, Rama Medical College (NCR), UP (India), as a randomized controlled trial and with parallel design during 2016-17. According to ADA [18], the whole study size was fifty patients with type 2 diabetic of mean age 52.3 ± 9.62 years were selected for the study. The sex, age, weight, height, duration of diabetes, blood pressure was examined ($\geq 130/\leq 80$ mm Hg) and recorded.

Blood samples (10 ml) were drawn from the patients and FBS, serum creatinine (SC), total cholesterol (TC), LDL, HDL, and HbA1c were measured before the initiation of supplementation with vitamin C and E. Subjects enrolled in the study received randomly 500 mg of vitamin C and 500 mg of vitamin E daily twice for 4 months. The patients were examined, and tests were repeated after the duration of 15 days of supplementation with vitamin C and E.

The blood sugar (FBS), creatinine, total cholesterol, high density lipoprotein (HDL), low density lipoprotein (LDL) and levels of HbA1c were measured by using Erba kits (Erba Lachema S.R.O., CZ) in the basal state and after the duration of 15 days of treatment. Venous blood (10 ml) was collected from each patient by a certified phlebotomist using standard laboratory methods at each study point.

In the study, Powers [19]; Barham and Trinder [20] proposed that, after clotting, blood was centrifuged at 2500 rpm for 30 minutes. The Serum glucose, creatinine (Myers et al. [21]), total cholesterol (Kannel et al. [22]), HDL (Castelli et al. [23]), LDL (Nauck et al. [24]) and HbA1c (Jeppsson et al. [25]) were assayed by colorimetric using Erba Reagent kits with EM 200-Automated Random Access Clinical Chemistry Analyzer [Erba Lachema S.R.O., CZ]. Total cholesterol, triglycerides, LDL, and HDL cholesterol were tested at baseline, subsequently monthly till next 4 months visits. Blood pressure was measured in the baseline and after every 15 days. It was measured in three positions (supine, sitting and upright) in 5 minutes intervals and the mean of them was calculated.

Statistical analysis was performed using Statistical Package for Social Sciences (SPSS 12.0, Chicago IL). Numerical normally distributed data and categorical data were compared using independent t-test. Results were given with their 95% CIs. Data were presented as means ± SD. Numerical normally distributed data and categorical data were compared using independent t-test.

According to Boshtam et al. [26], vitamin E could improve blood sugar in diabetic patients. According to Gazis et al. [27], the alpha tocopherol supplementation [1600 IU] in diabetic patients considerably reduced HbA1c levels. In the study, Paolisso et al. [28] proposed that, the HbA1c levels could be reduced by supplementation of 100 IU vitamin E in type 1 diabetic patients. Afkhami-Ardekani and Shojaoddiny-Ardekani [29] proposed significant decrease in FBS, TG, LDL, HbA1c and serum insulin was observed in diabetic patients supplemented with 1000 mg vitamin C.

Forghani et al. [30] proposed significant decrease in serum HbA1c and LDL levels observed in diabetic patients supplemented with 1000 mg/day of vitamin C for 6 weeks but according to Bishop et al. [31] supplementation only 500 mg/day vitamin C resulted in no significant changes observed in FBS, TC, TG and HbA1c level. Therefore, in present study, the combined doses of vitamin C and E [500 mg/ day twice] were supplemented to diabetic patients for 4 months and change in the level of FBS, serum creatinine (SC), total cholesterol (TC), HDL, LDL and levels of HbA1c were measured (Table 1). A significant decrease for FBS (from 162.78 ± 4.82 to 147.09 ± 7.05, p = 0. 00001) and total cholesterol (TC) [from 196.11 ± 9.83 to 188.75 ± 2.88, p = 0. 00001] was observed at 4 months in the 500 mg/day of vitamin C and E supplementation.

The multiple comparison analysis showed a borderline, not significant decrease in serum creatinine (SC) for the 500 mg group [vitamin C and E] at 4 months (from 1.32 ± 0.28 to 1.19 ± 0.13 mg/dl, P = 0. 00183), low density lipoprotein (LDL) was (from 128.02 ± 4.97 to 123.03 ± 1.40, p = < 0.00001) and in HbA1c level non-significant decline observed at 4 months (from 6.62 ± 2.404 to 5.14 ± 1.17, P = 0.00411). Non-significant raise observed in high density lipoprotein (HDL) [from 44.95 ± 8.13 to 47.83 ± 0.74, p = 0. 00802] and in calcium level (from 8.91 ± 0.438 to 9.35 ± 0.31, P = 0.00203) at 4 months in vitamin C and E [500 mg/day twice] supplemented diabetic group (Table 1, Figure 1). In the study, Afkhami-Ardekani [32] proposed that, the previous clinical trials showed a significant decrease in FBS, LDL and HbA1c levels after usage of 1000 mg of vitamin C or vitamin E separately in type 2 diabetic patients. According to Chen et al. [33] the daily consumption of 800 mg ascorbic acid for 4 weeks by type 2 diabetes patients caused no significant changes in FBS and serum insulin due to use of lower doses.

Forghani et al. [34] proposed significant decrease in serum HbA1c and LDL levels observed in patients supplemented with 1000 mg/day of vitamin C for 6 weeks. Errikson and Kahvakka [35] proposed significant decrease in TC was observed by using 2 gm of vitamin C for 90 days. In the study, Mullan [36] proposed that, vitamin C is required for regeneration of α-tocopherol and may thus prevent LDL oxidation in type 2 diabetes patients.

Paolisso et al. [37] proposed that, supplementation with 500 mg vitamin C twice daily for 4 months reduced the plasma levels of LDL, TC, TG and insulin significantly. In another study Watts et al. [38] proposed that, administration of 800 IU/day alpha tocopherol for 6 weeks has not beneficial effect on serum glucose and HbA1c in type 2 diabetic women. Cinaz et al. [39] observed that 900 IU/day vitamin E can improve insulin due to oxidative stress reduction. Manzella et al. [40] proposed that, 600 IU/day vitamin E supplementation reduced HbA1c, plasma insulin and oxidative stress indexes.

Table 2.1. The mean values of fasting blood sugar (FBS) serum creatinine (SC), total Cholesterol (TC) high density lipoprotein (HDL), low density lipoprotein (LDL), HbA1c (glycated haemoglobin) before and after supplementation with doses of vitamin C and E [(Data are mean ± SD)]

Variable	Control Group Range	Diabetic Group				p value
		Before Treatment	After Treatment			
		N = 50	N = 50			
		Mean ± SD	Mean ± SD	Variance (SD)	Population SD	
FBS (mg/dl)	70 - 110	162.78 ± 4.82	147.09 ± 7.05	49.73	6.10	0. 00001
SC (mg/dl)	0.6 – 1.2	1.32 ± 0.28	1.19 ± 0.13	0.02	0.11	0. 00183
TC (mg/dl)	185 - 190	196.11 ± 9.83	188.75 ± 2.88	8.31	2.49	0. 00001
HDL (mg/dl)	30 – 65	44.95 ± 8.13	47.83 ± 0.74	0.55	0.64	0. 00802
LDL (mg/dl)	80 - 150	128.02 ± 4.97	123.03 ± 1.40	1.96	1.21	< 0.00001
Calcium (mg/dl)	9 – 11	8.91 ± 0.438	9.35 ± 0.31	0.09	0.26	0.00203
HbA1C (mg/dl)	4 – 5.6	6.62 ± 2.404	5.14 ± 1.17	1.37	1.01	0.00411

$P* < 0.05 ** < 0.001$ compared to before treatment.

According to Paolisso and Giugliano [41] administration of vitamin E reduced triglycerides, total cholesterol and LDL. In another study Jain et al. [42] observed that, 100 IU/day vitamin E in diabetic patients reduced serum triglycerides significantly.

Cinaz et al. [43] and Boshtam et al. [44] did not show the effect of vitamin E on lipids. According to previous findings, a low dose of single vitamin not causes any significant benefits in diabetes patients. In present study, low combined doses of vitamin C and vitamin E [500 mg of both] twice daily for 4 months caused significant reduction in FBS, and TC level. There were no significant declines in SC, LDL and HbA1c level after supplementation of vitamin C and E [500 mg of both] twice daily for 4 months. A non-significant raise in HDL and calcium level was observed in patients with type 2 diabetes (Figure 2.1, Table 2.1).

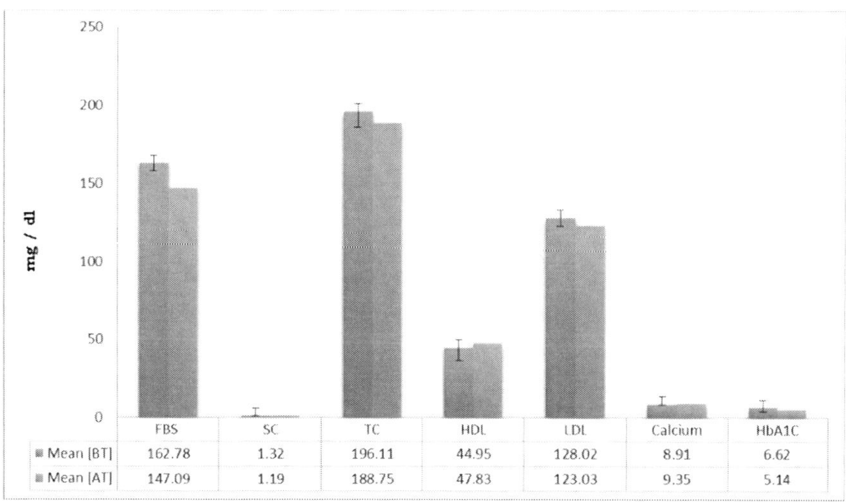

	FBS	SC	TC	HDL	LDL	Calcium	HbA1C
Mean [BT]	162.78	1.32	196.11	44.95	128.02	8.91	6.62
Mean [AT]	147.09	1.19	188.75	47.83	123.03	9.35	5.14

Figure 2.1. Standard Deviation graph showing mean values and/with positive, negative error of fasting blood sugar (FBS) serum creatinine (SC), total Cholesterol (TC) high density lipoprotein (HDL), low density lipoprotein (LDL), HbA1c (glycated haemoglobin) before [blue bars] and after supplementation [red bars] with doses of vitamin C and E [(Data are mean ± SD)].

CONCLUSION

In conclusion, supplementation of combined doses [500 mg/day twice] for 4 months, of vitamin C and E in addition to the normal diet may improve plasma glucose (FBS) and lipid profile in patients with type 2 diabetes. Overall, the significant reduction in FBS, and TC level was seen. There were no significant declines in SC, LDL and HbA1c level. A non-significant raise in HDL and calcium level was observed in patients with type 2 diabetes. Possibly this was due to insufficient samples, dosage or short duration of research. So, further studies with longer duration and higher dosage are suggested. The combined low doses of vitamin C and E twice in day may cause similar effect as caused by the higher doses of these vitamins individually.

REFERENCES

[1]   Pandit A, and Pandey AK. 2016. "Estimated Glomerular Filtration Rate and Associated Clinical and Biochemical Characteristics in Type 2 Diabetes Patients." *Adv in Diabetes and Metabol*. 4: 65 - 72. doi: 10.13189/adm.2016.040402.

[2]   Awasthi A, Parween N, Singh VK, Anwar A, Prasad B, and Kumar J. 2016. "Diabetes: Symptoms, Cause and Potential Natural Therapeutic Methods." *Adv. in Diabetes and Metabolism* 4, 10-23. doi: 10.13189/adm.2016.040102.

[3]   Prajapat R, and Bhattacharya I. 2016. "*In-silico* Structure Modelling and Docking Studies Using Dipeptidyl peptidase-4 (DPP4) Inhibitors against Diabetes Type-2." *Adv in Diabetes and Metabol*. 4: 73-84. doi: 10.13189/adm.2016.040403.

[4]   Zhaolan L, Chaowei F, Weibing W, and Biao X. 2010. "Prevalence of chronic complications of type 2 diabetes mellitus in outpatients-

a correstional hospital based survey in urban China." *HQLO* 8: 62-71.

[5]   Seyed, Hosseini SM, Mardanshahi A, Heidari SS, Sadr-Bafghi SMH, and Sadr-Bafghi SA. 2015. "Depression and Glycemic Control in Type II Diabetic Patients." *Iranian J of diabetes and obesity* 7 (3):112-117.

[6]   Rahman, Hassan SAEL, Rahman, Elsheikh WA, Abdel, Rahman NI, Nabiela, ElBagir M. 2016. "Serum Calcium Levels in Correlation with Glycated Hemoglobin in Type 2 Diabetic Sudanese Patients." *Adv in Diabetes and Metabolism* 4: 59 - 64. doi: 10.13189/adm.2016.040401.

[7]   Khabaz M, Rashidi M, Kaseb F, and Afkhami-Ardekan M. "Effect of Vitamin E on Blood Glucose, Lipid Profile and Blood Pressure in Type 2 Diabetic Patients. *Iranian J of Diabetes and Obesity* 1 (1): 10-15.

[8]   Manzella D, Barbieri M, Ragno E, and Paolisso G. 2001. "Chronic administration of harmacologic doses of vitamin E improves the cardiac autonomic nervous system in patients with type 2 diabetes." *Am J Clin Nutr.* 73: 1052–1057.

[9]   Afkhami-Ardekani M, Vahidi AR, Borjian L, and Borjian L. 2003. "Effect of vitamin C supplement on glycosylated hemoglobin in patients with type 2 diabetes." *J Shah Sad Univ.* 10:15-8.

[10] Dakhale GN, Chaudhari HV, and Shrivastava M. 2011. "Supplementation of Vitamin C Reduces Blood Glucose and Improves Glycosylated Hemoglobin in Type 2 Diabetes Mellitus: A Randomized, Double-Blind Study." *Adv in Pharmacological Sciences*, vol. 2011, doi:10.1155/2011/195 271

[11] Chambial S, Dwivedi S, Shukla KK, John PJ, and Sharma P. 2013. "Vitamin C in Disease Prevention and Cure: An Overview." *Indian J Clin Biochem.* 28 (4): 314–328. doi: 10.1007/s12291-013-0375-3.

[12] Simom JA. 1992. "Vitamin C and cardiovascular disease: a review." *J Am Coll Nutr.* 11: 107-25.

[13] Ness AR, Khaw KT, Bingham S, and Day NE. "Vitamin C status and serum lipids." *Eur J Clin Nutr* 5: 724-9.

[14] Battisi WP, Palmisano J, and Keane WE. 2003. "Dyslipidemia in patients with type 2 diabetes: Relation between lipids, kidney disease and cardiovascular disease." *Clin Chem Lab Med.* 41: 881-91.

[15] Manjunath CN, Rawal JR, Irani PM, and Madhu K. 2013. "Atherogenic dyslipidemia." *Indian J Endocrinol Metab.* 17(6): 969–976. doi: 10.4103/2230-8210.122600

[16] Errikson J, and Kahvakka A. 1995. "Magnesium and ascorbic acid supplementation in diabetes mellitus." *Ann Nutr Metab.* 39: 217-23.

[17] Sargeant LA, Wareham NJ, Bingham Luben RN, Oakes S, and Welch A, et al., 2000. "Vitamin C and hyperglycemia in the European prospective investigation into cancer- Norfolk (EPIC-Norfolk) study." *Diabetes Care.* 23: 726-32.

[18] American Diabetes Association. Standards of medical care in diabetes-2006. *Diabetes Care.* 2006; 29:4-42.

[19] Powers AC. 2001. *Diabetes mellitus. Harrison's principles of internal medicine.* New York: McGraw Hill : 2124-5.

[20] Barham D, and Trinder P. 1972. "An improved color reagent for the determination of blood glukose by the oxidase system." *Analyst.* 97: 142 - 5.

[21] Myers GL, Miller WG, Coresh J, Fleming J, Greenberg N, Greene T, Hostetter T, Andrew SL, Panteghini M, Welch M, and Eckfeldt JH. 2006. "Recommendations for Improving Serum Creatinine Measurement: A report from laboratory working group of the National kidney disease education program." *Clinical Chemistry* 52 (1): 5 – 18.

[22] Kannel WB, Castelli WP, Gordon T. 1979. "Cholesterol in the diction of atherosclerotic disease; new perspectives based on the Framingham study." *Ann Intern Med.* 90:85.

In: Medical Bioinformatics and Biochemistry   ISBN: 978-1-53614-952-4
Editors: Rajneesh Prajapat et al.   © 2019 Nova Science Publishers, Inc.

Chapter 3

IN-SILICO STRUCTURE MODELLING AND DOCKING STUDIES USING DIPEPTIDYL PEPTIDASE-4 (DPP4) INHIBITORS AGAINST DIABETES TYPE-2

Rajneesh Prajapat[*] *and Ijen Bhattacharya*
Department of Medical Biochemistry,
Faculty of Medical Sciences, Rama Medical College
and Hospital (NCR), Rama University, Kanpur, UP, India

ABSTRACT

Recently recognized class of oral hypoglycemic, dipeptidyl peptidase (DPP4) inhibitors could block the dipeptidyl peptidase-4 (DPP4) enzymes. DPP4 is an intrinsic membrane glycoprotein and a serine exopeptidase that plays a major role in glucose metabolism and responsible for the degradation of incretins such as GLP-1, therefore providing a useful treatment to diabetes mellitus type 2. The present work focused on the

---

[*] Corresponding Author's Email: prajapat.rajneesh@gmail.com.

study of the structural homology modelling of dipeptidyl peptidase-4 [*Homo sapiens*] (NP_001926). The Ramachandran plot of DPP4 (NP_001926.2) has 88.9% residues in the most favored region while template 2QT9 has 96.1% residues in the most favored region. The model was validated by using protein structure tools RAMPAGE and Prochek for reliability. Docking studies were further performed to analyze the interaction mode between selected DPP4 inhibitor anagliptin derivative SKK and receptor DPP4 by using Hex 8.0.0. The *in silico* analysis was useful to identify the novel inhibitor that illustrate better activity than the other reported inhibitors.

Keywords: *In silico*, DPP4, GLP-1, Hex

3. INTRODUCTION

Diabetes mellitus is a prime public health problem and forthcoming epidemic all over the globe [38] disorder caused due to insufficient or ineffective insulin [2]. In India, there were approximately 40 million people suffering from diabetes and this number possibly will rise to 300 million by 2025 [45, 46]. According to International Diabetes Federation record, worldwide the number of people with diabetes will be increase from current figure of 240 million to 380 million over the next 20 years [11, 37] and up to 642 million by 2040 [10].

The combination therapeutics or oral monotherapy with other anti-diabetic agents used to control diabetes through clinical anti-diabetes therapy [16]. But, these anti-diabetic agents may cause adverse side effects and chronic complications [6, 25].

Dipeptidyl peptidase-4 (DPP4) encoded by the DPP4 gene and it is a membrane serine exopeptidase that cleaves X-proline dipeptides from the N-terminus of polypeptides [12, 40]. DPP-4 inhibitors recommended for use in patients with T2DM therapy, that including hypoglycaemia incidences, risks of cardiovascular complications and weight gain [14]. It is a rather indiscriminate enzyme for which a diverse range of substrates are known. Dipeptidyl peptidase-4 (DPP4) is a serine protease that [36] that degrade peptide hormone glucagon-like peptide 1 (GLP-1). GLP-1 plays an important role in the regulation of insulin release to control the level of blood sugar in human body [18]. Several studies demonstrated that the inhibition of DPP4 could increase the amount of circulating GLP-1 to improve the secretion of insulin in the body [43]. Therefore, it has regarded as promising target to develop novel drug for treatment of type 2 diabetes.

So far, a couple of identified DPP4 inhibitors, such as sitagliptin, were approved to be used clinically as antidiabetic drugs by FDA [41; 20]. However, there is still a need for more potent, selective and safer DPP4 inhibitor, which does not have the in specificity and side effect possessed by the presently available inhibitors [13], because of worldwide problem of type 2 diabetes. The recent research on DPP4 inhibitors using *in silico* methods is focus on physicochemical analysis [3]. Therefore, researcher is studying new DPP4 ligands for the development of the novel anti-diabetes drug.

Homology modelling provides structural information about the protein dynamics, function, interactions with other proteins and ligands [7, 31]. In the present study, homology modelling [22] process was used to determine the structure of dipeptidyl peptidase-4 [*Homo sapiens*] (NP_001926).

The docking was performed to reveal the interaction mode and to predict the orientation [29, 30] between inhibitors SKK and DPP4 by using Hex 8.0.0. The docking outcome demonstrated the binding conformation of anagliptin derivative SKK and DPP4. The results of the study could be used to design or predict new potent DPP4 inhibitors. *In silico* analysis is an effective way to develop the anti-diabetic drug model.

3.1. METHODOLOGY

3.1.1. Sequences Retrieval and Alignment

The amino acid sequence of dipeptidyl peptidase-4 [*Homo sapiens*] (NP_001926) was retrieved from NCBI (www.ncbi.nlm.nih.gov). Multiple sequence alignment was performed by using BLASTp [1; 44] against the PDB (Protein Databank), to find out the related homologues. The PDB file of DPP4 (NP_001926) was generated by using 3D-JIGSAW. The PDB file of homologous template and query were further utilized for 3D model energy validation [17].

3.1.2. Molecular Modelling of DPP4

The UCLA-DOE provide an analysis about the quality of a putative crystal structure for protein. RAMPAGE program was used to visualize and evaluate the Ramachandran plot of DPP4 (NP_001926) and its homologous template [23]. The models were analyzed based on various factors such as G-factor, number of residues in core, generously allowed and disallowed regions in Ramachandran plot. The validation of structure models was performed by using PROCHECK [21]. QMEAN [8] and ProSA [42] were further used for the analysis of model. The ProSA server displays the Z-score and energy plots.

The 3D structure selected as a template for constructing the model for DPP4 (NP_001926) was Human dipeptidyl peptidase (PDB access code: 2QT9). 2QT9 has 766 amino acid residues and a resolution of 2.1 Å [19]. 2QT9 was illustrated 99% identity with the target sequence (2QT9). The quality of folding was checked by using Verify3D. The system energy and 3D alignment were monitored using Atomic

Non-Local Environment Assessment (ANOLEA) [24]. Produced models of homologous template and query were ranked on QMEAN server and utilized for the ribbon structural model construction. QMEAN locate the position of active site amino acids of protein and estimates the per-residue error [9].

3.1.3. Docking of DPP4 (NP_001926) with SKK

The aim of protein docking is to determine 3D structure of protein complex preliminary from its unbound components [15]. Docking was performed to analyze the interaction between selected DPP4 inhibitor anagliptin derivative SKK [Synonyms: Anagliptin, $C_{19}H_{25}N_7O_2$] (Figure 3.1) and receptor DPP4 (NP_001926) by using Hex 8.0.0 [34]. The PDB files of DPP4 (NP_001926) and SKK was used as inputs for the protein-protein docking. Hex 8.0.0 is a docking program; that performed structural refinement, energy minimization and display the fast 3D superposition using the SPF correlation approach. Based on the energy minimization the best front of the docked complex was selected [35].

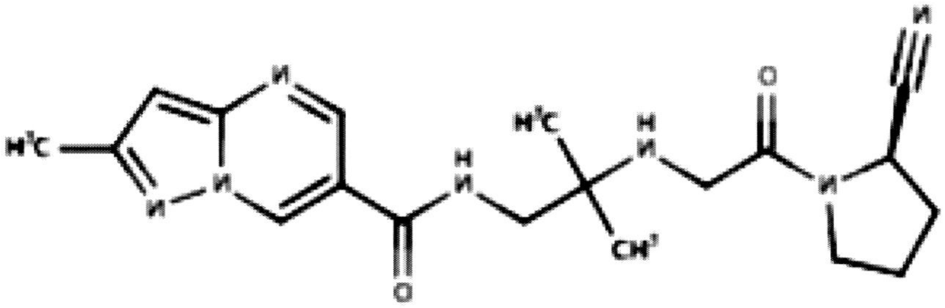

Figure 3.1. SKK (N-[2-({2-[(2S)-2-cyanopyrrolidin-1-yl]-2-oxoethyl}amino)-2-methylpropyl]-2-methylpyrazolo[1,5-a]pyrimidine-6-carboxamide).

3.2. RESULTS AND DISCUSSION

3.2.1. Building of Protein Model

The sequence alignment of dipeptidyl peptidase-4 [*Homo sapiens*] (NP_001926) revealed the sequence homology with human dipeptidyl peptidase (PDB access code: 2QT9), which has 766 amino acid residues and a resolution of 2.1 Å [19]. The 2QT9 (with 99% sequence identity) was selected as template for the model building of DPP4 (NP_001926) protein. Total 766 residues were modeled with 99% confidence by the single highest scoring template. To build the model, BLAST was run with the maximum E-value allowed for template being 0.0.

3.2.2. Model Reputation

The DPP4 (NP_001926) model was illustrated good stereo-chemical property in terms of overall G-factor value of -0.73. The probability conformation with 88.9% residues in the favored region of Ramachandran plot, illustrated high accuracy of model predicted [32]. The residues number in allowed and outlier region of plot were 8.0% and 19% (Figure 3.2). The Ramachandran plot of template (2QT9) has 96.1% residues in favored region, 3.9% in allowed region and 0.1% in outlier regions. (Figure 3.3, Table 3.1).

Table 3.1. Results summary of the Ramachandran plot

Accession No	Protein	Description	Residues (%)		
			Favored regions	Allowed regions	Outlier regions
NP_001926.2	DPP4	*Homo sapiens*	88.9	8.0	19%
2QT9	DPP IvCD26	*Homo sapiens*	96.1	3.9	0.1

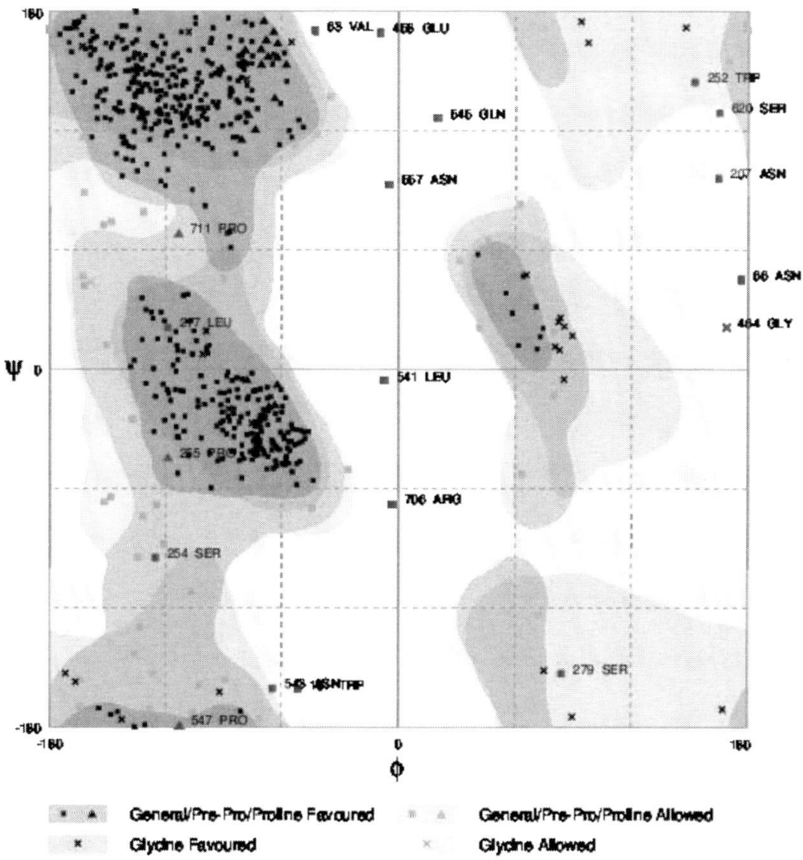

Figure 3.2a. Ramachandran Plot analysis of dipeptidyl peptidase-4 (NP_001926.2) [*Homo sapiens*] protein. (a) Total number of residues were 601 with 88.9% in most favored regions [A, B, L], 8.0 % in allowed regions [a,b,l,p], and 19% in outlier region regions.

The Ramachandran plot of DPP4 (NP_001926) has only 88.9% residues in the most favored region and its template 2QT9 [Human dipeptidyl peptidase] has higher 96.1% residues in most favored regions, hence 2QT9 is more stable than DPP4 (NP_001926) (Table 1). A good quality Ramachandran plot has more than 90% in the most favored regions [26] thus the 2QT9 model could be included in good quality category. For target DPP4 (NP_001926) model, energy minimization

 Rajneesh Prajapat and Ijen Bhattacharya

should be required to enhance stability by using application of simulated annealing, the steepest descent and conjugate gradient.

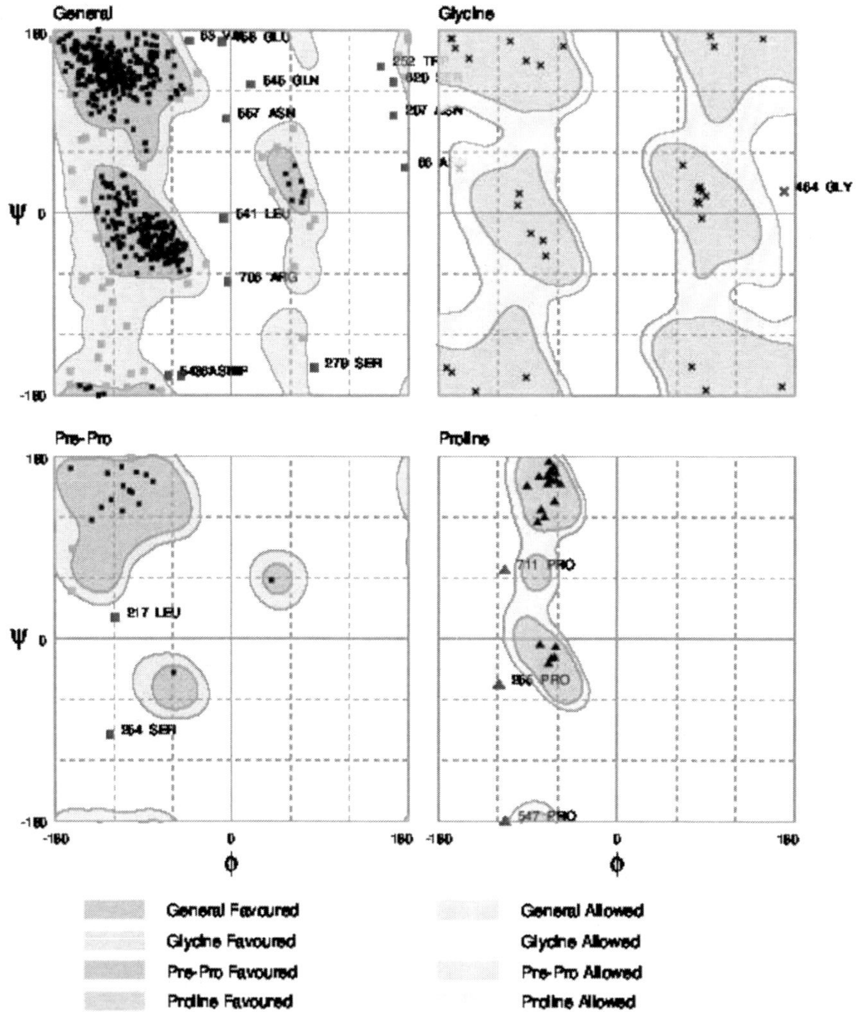

Figure 3.2b. Non-proline residues and non-glycine residue regions.

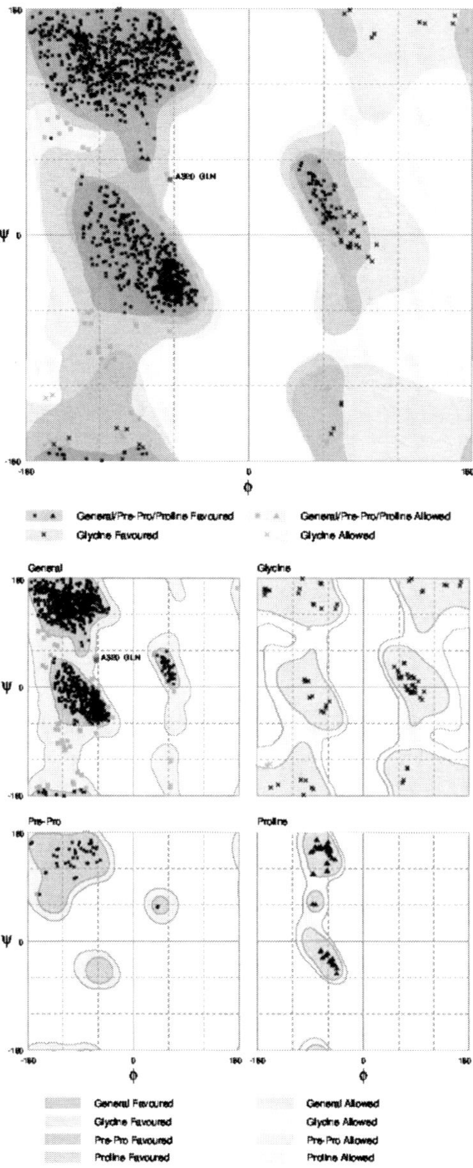

Figure 3.3. Ramachandran Plot analysis of homologous Human Dipeptidyl Peptidase IvCD26 in complex with a 4-Aryl Cyclohexylalanine Inhibitor (2QT9) protein. (a) Total number of residues were 1452 with 96.1% in most favored regions [A, B, L], 3.9 % in allowed regions [a,b,l,p], and 0.1% in outlier region regions. (b) Non-proline residues and non-glycine residue regions.

3.2.3. Model Validation

The problems of protein structure based on energy plots could be easily seen by ProSA and displayed in a three-dimensional manner. ProSA web was used to check the three-dimensional model errors of DPP4 (NP_001926) and 2QT9 (Figure 3.4 and Figure 3.5). ProSA web z-scores of protein chains were determined by X-ray crystallography (light blue) or NMR spectroscopy (dark blue) with respect to their length. The z-scores of DPP4 (NP_001926) and DPP IvCD26 (2QT9) is highlighted as large dots.

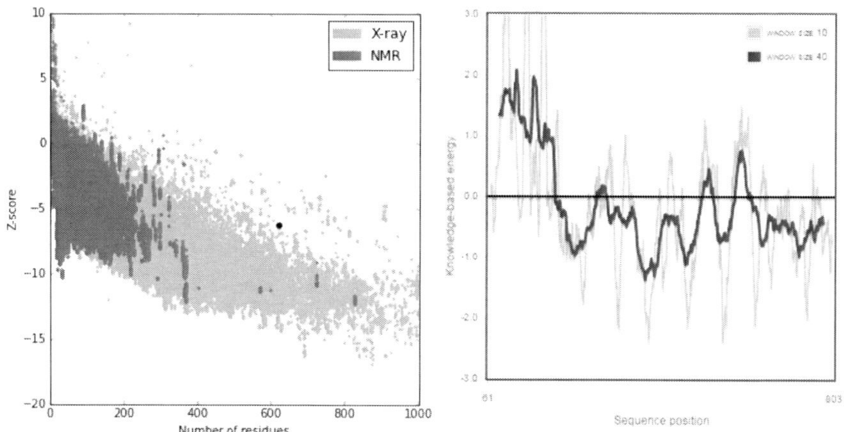

Figure 3.4. ProSA web service analysis of DPP4 (NP_001926) [*Homo sapiens*] protein overall model quality (a) and local model quality (b).

The ProSA z-score was -6.21 for DPP4 (NP_001926) [Figure 4] and -11.09 for the DPP IvCD26 (2QT9) protein, indicates the overall model quality of target and template (Figure 3.5) measures the deviation of the total energy of the structure with respect to an energy distribution [39].

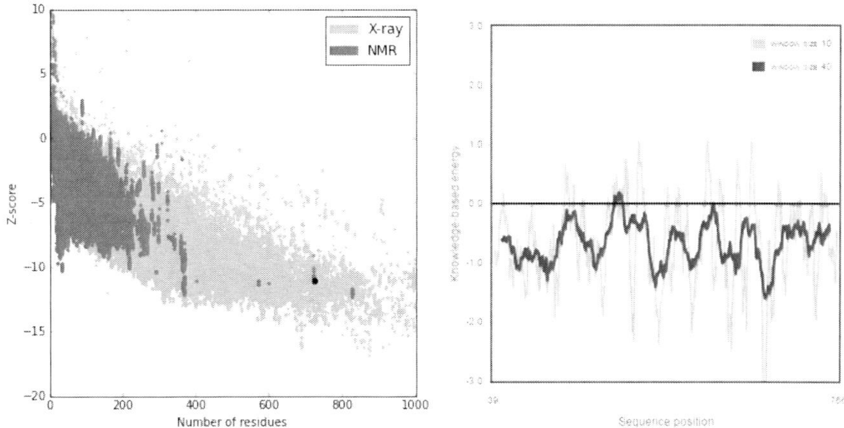

Figure 3.5. ProSA web service analysis of DPP IvCD26 (2QT9) [*Homo sapiens*] protein overall model quality (a) and local model quality (b).

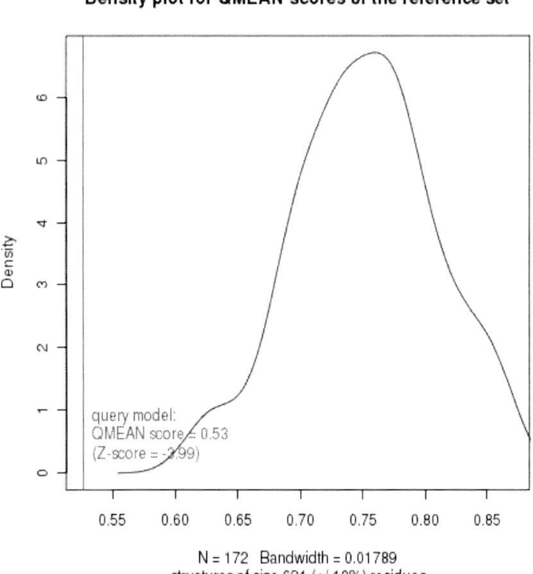

Figure 3.6a. The density plot for target DPP4 (NP_001926) [*Homo sapiens*] showing the value of Z-score and QMEAN score.

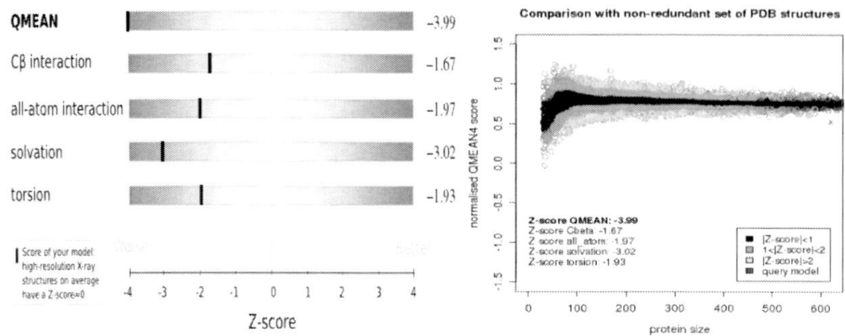

Figure 3.6b. Plot showing the QMEAN value as well as Z-score (for NP_001926).

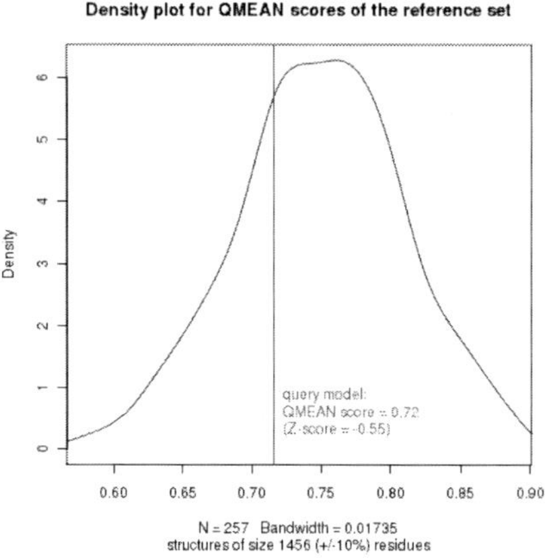

Figure 3.7a. The density plot for target DPP IvCD26 (2QT9) [*Homo sapiens*] showing the value of Z-score and QMEAN score.

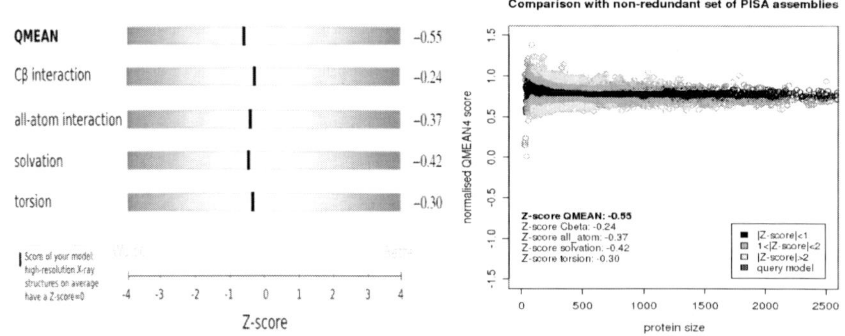

Figure 3.7b. Plot showing the QMEAN value as well as Z-score (for 2QT9).

QMEAN score of target DPP4 (NP_001926) model was 0.53 and Z-score was -3.99 (Figure 6a). QMEAN score of template DPP IvCD26 (2QT9) was 0.72 with Z-score was -0.55 (Figure 7a), which very closed to the value of 0. That illustrated the fine quality of the both models [42]. The reliability of the model was expected in between 0 and 1 and this could be inferred from the density plot for QMEAN scores of the reference set. A comparison between protein size and normalized QMEAN score (0.40) in non-redundant set of PDB structures plot revealed different set of z-values for different parameters such as C-beta interactions (-1.67), interactions between all atoms (-1.97), solvation (-3.02) and torsion (-1.93) [Figure 6b]. The ribbon model of DPP4 (NP_001926) and 2QT9 was generated using QMEAN server (Figure 8 and Figure 9). The per-residue error visualized using a color gradient from blue (more reliable region) to red (potentially unreliable region).

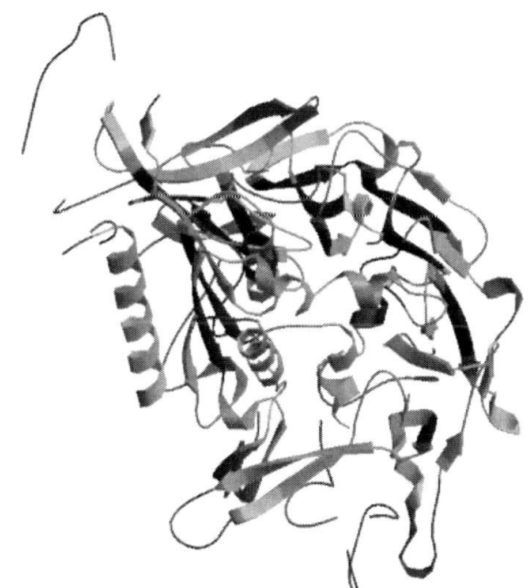

Figure 3.8. DPP4 (NP_001926) [*Homo sapiens*] ribbon model generated using QMEAN server. Estimated per-residue error visualized using a color gradient from blue (more reliable region) to red (potentially unreliable region).

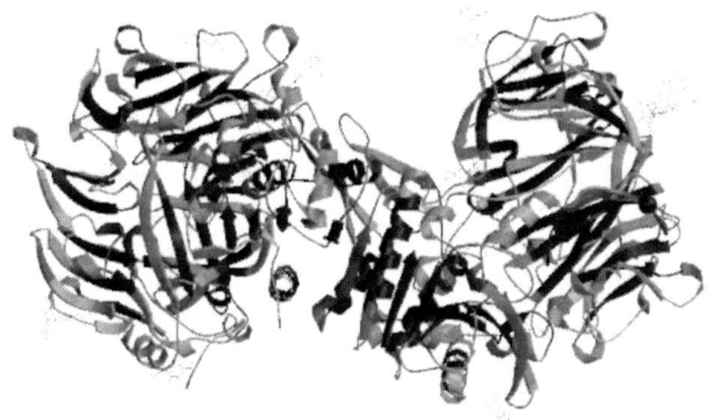

Figure 3.9. DPP IvCD26 (2QT9) [*Homo sapiens*] ribbon model generated using QMEAN server. Estimated per-residue error visualized using a color gradient from blue (more reliable region) to red (potentially unreliable region).

Protein docking predicts symmetrical 3D models of protein-protein complexes with arbitrary point group symmetry [33]. Hex assigns multiple local coordinate systems to larger molecule (receptor DPP4 [NP_001926]) and docks the ligand (SKK) around each local coordinate frame on the receptor. The figure represents Hex view of DPP4 (NP_001926) protein domain and SKK in Van der Waals mode, with the inter-molecular axis (Figure 10a) and solid surfaces form (Figure 10b).

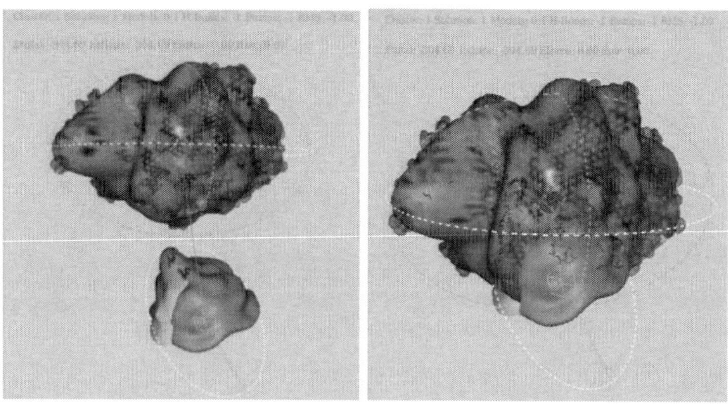

Figure 3.10. A Hex scene showing the DPP4 (NP_001926) domain and SKK (below) in Vander Waals mode and with the intermolecular axis drawn in blue (a) and solid surfaces (b).

Table 3.2. Clustering found 560 clusters from 2000 docking solutions in 0.05 seconds

Clst	Soln	Models	E-total	E-shape	E-force	E-air	Bmp	RMS
1	1	000:001	-304.7	-304.7	0.00	0.0	-1	-1.00
1	2	000:001	-304.6	-304.6	0.00	0.0	-1	-1.00
1	7	000:001	-295.7	-295.7	0.00	0.0	-1	-1.00
1	8	000:001	-294.7	-295.7	0.00	0.0	-1	-1.00
1	9	000:002	-292.7	-292.7	0.00	0.0	-1	-1.00
1	11	000:001	-288.3	-288.3	0.00	0.0	-1	-1.00
1	12	000:002	-286.3	-286.3	0.00	0.0	-1	-1.00
1	21	000:002	-282.4	-282.4	0.00	0.0	-1	-1.00
1	23	000:001	-280.5	-280.5	0.00	0.0	-1	-1.00
1	28	000:002	-277.7	-277.7	0.00	0.0	-1	-1.00
1	30	000:001	-276.0	-276.0	0.00	0.0	-1	-1.00
1	37	000:001	-272.6	-272.6	0.00	0.0	-1	-1.00
1	43	000:002	-270.8	-270.8	0.00	0.0	-1	-1.00
1	44	000:002	-270.6	-270.6	0.00	0.0	-1	-1.00
1	45	000:001	-270.4	-270.4	0.00	0.0	-1	-1.00
1	49	000:001	-270.2	-270.2	0.00	0.0	-1	-1.00
1	54	000:001	-268.8	-268.8	0.00	0.0	-1	-1.00
1	65	000:001	-265.7	-265.7	0.00	0.0	-1	-1.00
1	78	000:001	-263.4	-263.4	0.00	0.0	-1	-1.00
1	87	000:001	-261.6	-261.6	0.00	0.0	-1	-1.00

Based on the RMS and energy values, the best docking orientation was selected. The better RMS value of docking was -1.00. The binding sites of protein reveal chemical specificity and determine the nature of ligand [4, 5]. The 1 Clst initial Etotal, Eshape and Eforce values for the model were -304.7, -304.7 and 0.0 (Table 3.2).

Figure illustrates spherical harmonic surfaces to order L = 12 for the DPP4 (NP_001926) protein domain and SKK (Figure 3.11). Illustration of the DPP4/SKK complex shown as contoured Gaussian density surfaces and background modes (Figure 3.11). Figure illustrates docking control results in the form of solid models (Figure 3.12a) and solid surface (Figure 3.12b) model view of DPP4 (NP_001926) and SKK complex. These docking results suggest that SKK interacts with the

      Rajneesh Prajapat and Ijen Bhattacharya

DPP4 protein of and causes its inhibition [28]. The binding pocket values for DPP4 model were predicted by using Hex 8.0.0. The predicted two pockets by the software with different primary surface area and volume shown in Table 3.3.

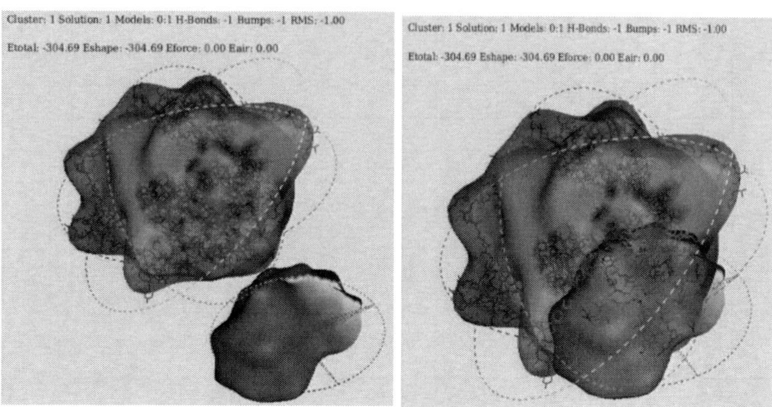

Figure 3.11. Illustration of spherical harmonic surfaces to order L = 12 for the DPP4 (NP_001926) domain and SKK (a). Illustration of the DPP4 (NP_001926) (Receptor) and SKK (Ligand) complex shown as contoured Gaussian density surfaces and colored by chain color, drawn using perspective (keyboard P) and background (keyboard B) modes (b).

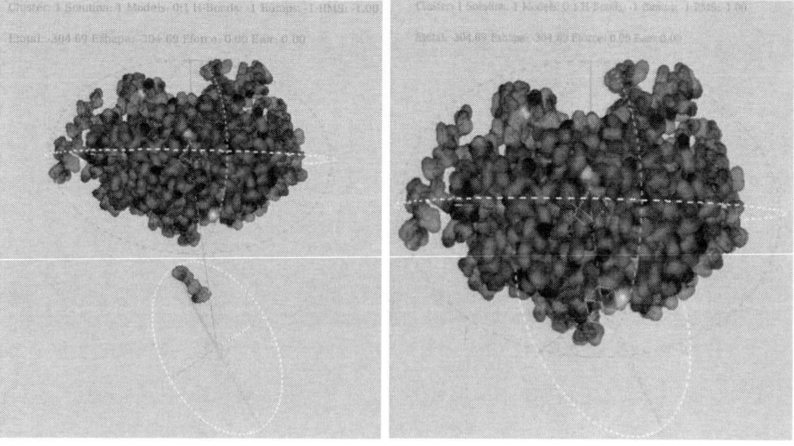

Figure 3.12. Docking results illustrate solid models (a) and solid surface (b) view forms of DPP4 (NP_001926) /SKK complex.

Table 3.3. Binding site model DPP4 (NP_001926) protein. The calculated 29 surface normal (docking orientations) in 0.28 seconds. Calculating canonical orientation to L = 12

Pocket	Polar Probe	Apolar probe	Primary Surface area	Primary Surface volume	Typical Edge arc	Typical edge length (Å)	Average Radius (Å)	Surface area (Å)	Triangles		
									Min	Max	Avg
1	0.00A	0.00A	51389.81	62777.83	4.62	2.59	32.14	20349.10	1.84	58.99	4.52
2	0.00A	0.00A	51084.10	66988.55	2.80	0.69	14.07	1.3433	0.00	37.04	1.24

Docking could be used to study the mechanism of enzymatic reaction, to identify possible binding modes for a ligand and to screen a database [27]. A few amino acids were found to be conserved in DPP4 (NP_001926), that forming the binding cavity for the SKK. Interaction of SKK with DPP4 (NP_001926), used as a guideline to design and predict new potent DPP4 inhibitors, which could be an effective way to find novel leads for the development of anti-diabetic drugs. The anaglyptic derivative improves the glycemic control; hence, SKK might be considered as novel potent DPP4 inhibitor.

CONCLUSION

The DPP4 inhibitors accepted as new drug class because it provides effective glycemic control and low risk of hypoglycemia. *In silico* Modelling and docking analyses illustrated that the anagliptin derivative SKK could be inhibited DPP4 enzymatic activity. The results of study could be used to design and predict new potent DPP4 inhibitors, which could be an efficient mode for designing of anti-diabetic drugs. The results will be further supportive to screen new DPP4 inhibitors from nature sources also. The *in silico* techniques are useful to identify the novel inhibitors for the clinical trial of diabetic patients.

REFERENCES

[1]   Altschul SF, Madden TL, Schaffer AA, Zhang J, Zhang Z, Miller W, and Lipman DJ. 1997. "Gapped BLAST and PSI-BLAST: A new generation of protein database search programs." *Nucleic Acids Res.* 25: 3389-3402.

[2]   Awasthi A, Parween N, Singh VK, Anwar A, Prasad B, and Kumar J. 2016. "Diabetes: Symptoms, Cause and Potential Natural Therapeutic Methods." *Adv. in Diabetes and Metabolism* 4: 10-23. doi: 10.13189/adm.2016.040102.

[3]   Balajee R, and Dhanarajan MS. 2012. "Identification and Comparative Molecular Docking Analysis of 6, 7, 8, 9-Tetrahydro-2*H*-11-oxa-2, 4, 10-triaza-benzo [B] fluoren-1-one (RBMS-01) Bounds with DPP4 for Anti-Hyperglycemic Activities." *Chem Sci Trans.* 1: 279-288. DOI:10.7598/cst2012.166.

[4]   Balakrishnan M, and Srivastava RC. 2009. "Protein Structural Modelling of Acetylglutamate Kinase from Leptospira Interrogans and Docking Studies of N-Acetyl-L-Glutamate." *The IUP J. of Biotech.* 3: 7-16.

[5]   Balakrishnan M, Srivastava RC, and Pokhriyal M. 2010. Homology modelling and docking studies between HIV-1 protease and carbamic acid. *Indian J. Biotechnol.* 9: 96-100.

[6]   Barnett AH. 2008. "Thiazolidinediones and cardiovascular outcomes." *Brit. J. Diabetes Vasc Dis.* 8:45-9.

[7]   Barry MM, Mol CD, Anderson WF, and Lee JS. 1994. "Sequencing and Modelling of anti-DNA immunoglobulin Fv domains. Comparison with crystal structures." *J. Biol Chem.* 269: 3623-3632.

[8]   Benkert P, Künzli M, and Schwede T. 2009. "QMEAN server for protein model quality estimation." *Nucleic Acids Res.* 37: W510-4. doi: 10.1093/nar/gkp322.

[9]   Benkert P, Schwede T, and Tosatto SCE. 2009. "QMEANclust: Estimation of protein model quality by combining a composite

scoring function with structural density information." *BMC Struct Biol.* 9:35. doi: 10.1186/1472-6807-9-35.

[10] Cefalu WT, Buse JB, Tuomilehto J, Fleming GA, Ferrannini E, Gerstein HC, Bennett PH, Ramachandran A, Raz I, Rosenstock J, and Kahn SE. 2016. "Update and next steps for Real-World Translation of Interventions for Type 2 Diabetes Prevention: Reflections from a Diabetes Care Editors' Expert Forum." *Diabetes Care* 39: 1186-1201. DOI: 10.2337/dc16-0873.

[11] Chada RR, Sethi BK, Waghray K, and Naidu SK. 2014. "Obesity and Type 2 Diabetes: A Population Based Study of Urban School Children in South India." *Adv. Diabetes and Metabolism* 2, 4-9. doi: 10.13189/adm.2014.020102.

[12] Darmoul D, Lacasa M, Baricault L, Marguet D, Sapin C, Trotot P, Barbat A, and Trugnan G. 1992. "Dipeptidyl peptidase IV (CD26) gene expression in enterocyte-like colon cancer cell lines HT-29 and Caco-2: cloning of the complete human coding sequence and changes of dipeptidyl peptidase IV mRNA levels during cell differentiation." *J. Biol Chem.* 267: 4824-4833.

[13] Defronzo RA, Okerson T, Viswanathan P, Guan X, Holcombe JH, and MacConell L. 2008. "Effects of exenatide versus sitagliptin on postprandial glucose, insulin and glucagon secretion, gastric emptying, and caloric intake: a randomized, cross-over study." *Curr Med Res Opin.* 24: 2943-2952.

[14] Galstyan KO, Nedosugova LV, Petunina NA, Trakhtenberg JA, Vostokova NV, Karavaeva OV, and Chasovskaya TE. 2015. "First Russian DPP-4 inhibitor Gosogliptin comparing to Vildagliptin in type 2 diabetes mellitus patients." *Diabetes mellitus* 19(1):89-96.

[15] Ghoorah AW, Devignes MD, Smaïl-Tabbone M, Ritchie DW. 2013. "Protein docking using case-based reasoning." *Proteins* 81: 2150-2158. doi:10.1002/prot.24433.

[16] Havale SH, and Pal M. 2009. "Medicinal chemistry approaches to the inhibition of dipeptidyl peptidase-4 for the treatment of type 2 diabetes." *Bioorg Med. Chem.* 17:1783–802.

[17] Heinrichs A. 2008. "Proteomics: Solving a 3D jigsaw puzzle." *Nat Rev Mol Cell Biol*. 9: 3-3.

[18] Jiang C, Han S, Chen T, and Chen J. 2012. "3D-QSAR and docking studies of arylmethylamine-based DPP IV inhibitors." *Acta Pharmaceutica Sinica B*. doi:10.1016/j.apsb.2012.06.007.

[19] Kaelin DE, Smenton AL, Eiermann GJ, He H, Leiting B, Lyons KA, Patel RA, Patel SB, Petrov A, Scapin G, Wu JK, Thornberry NA, Weber AE, and Duffy JL. 2007. "4-arylcyclohexylalanine analogs as potent, selective, and orally active inhibitors of dipeptidyl peptidase IV." *Bioorg Med Chem Lett*. 17:5806-11.

[20] Kim D, Kowalchick JE, Edmondson SD, Mastracchio A, Xu J, and Eiermann GJ et al., 2007. "Triazolopiperazine-amides as dipeptidyl peptidase IV inhibitors: close analogs of JANUVIA (sitagliptin phosphate)." *Bioorg Med Chem Lett*. 17: 3373-3377.

[21] Laskowski RA, Rullmann JA, MacArthur MW, Kaptein R, and Thornton JM. 1996. "AQUA and PROCHECK-NMR: Programs for checking the quality of protein structures solved by NMR." *J. Biomol. NMR*. 8: 477-486.

[22] Lima AH, Souza PRM, Alencar N, Lameira J, Govender T, Kruger HG, Maguire GEM, and Alves CN. 2012. "Molecular Modelling of *T. rangeli*, *T. brucei gambiense*, and *T. evansi* Sialidases in complex with the DANA inhibitor." *Chem Biol Drug Des*. 80: 114-120.

[23] Lovell SC, Davis IW, Arendall WB III, Bakker de PIW, Word JM, Prisant MG, Richardson JS, and Richardson DC. 2002. "Structure validation by Calpha geometry: phi, psi and Cbeta deviation." *Proteins: Structure, Function & Genetics* 50: 437-450.

[24] Melo F, Devos D, Depiereux E, and Feytmans E. 1997. "ANOLEA: a www server to assess protein structures." *Proc. Int. Conf. Intell. Syst. Mol. Biol*. 5: 187-190.

[25] Mizun CS, Chittiboyina AG, Kurtz TW, Pershadsingh HA, and Avery MA. 2008. "Type 2 diabetes and oral anti hyperglycemic drugs." *Curr Med Chem*. 15:61-74.

[26] Morris AL, MacArthur MW, Hutchinson EG, and Thornton JM. 1992. "Stereochemical quality of protein structure coordinates." *Proteins* 12: 345-364.

[27] Morris GM, and Lim-Wilby M. 2008. "Molecular Docking, Methods in Molecular Biology." In: *Molecular Modelling of Proteins*, Kukol, A. (Ed.). Humana Press, Totowa, New Jersey, pp: 365-382.

[28] Mustard D, and Ritchie DW. 2005. "Docking essential Dynamics Eigenstructures." *Proteins: Struct Funct Bioinfo.* 60: 269-274.

[29] Nongonierma AB, and FitzGerald RJ. 2014. "An *in silico* model to predict the potential of dietary proteins as sources of dipeptidyl peptidase IV (DPP-IV) inhibitory peptides." *Food Chemistry* 165: 489-498.

[30] Nukala UA, Sahithi P, and Rao PR. 2015. "*In-silico* Structure based, QSAR and Analogue based Studies using Dipeptidyl peptidase-4 (DPP4) Inhibitors against Diabetes Type-2." *Int J. of Biote & Bio Sci.* 1 (1): 16-20.

[31] Prajapat R, and Bhattacharya I. 2016. "*In Silico* Structure Analysis of Type 2 Diabetes Associated Cysteine Protease Calpain-10 (CAPN10)." *Adv in Diabetes and Metabolism* 4: 32-43. doi: 10.13189/adm.2016.040202.

[32] Prajapat R, Marwal A, and Gaur RK. 2014. "Recognition of Errors in the Refinement and validation of three-dimensional structures of AC1 proteins of begomovirus strains by using ProSA-Web." *J. of Viruses* ID 752656, 6 pages.doi.org/10.1155/2014/752656.

[33] Ritchie DW, and Grudinin S. 2016. "Spherical Polar Fourier Assembly of Protein Complexes with Arbitrary Point Group Symmetry." *J. of Applied Crystallography* 49: 158-167.

[34] Ritchie DW, Kozakov D, and Vajda S. 2008. "Accelerating protein-protein docking correlations using a six-dimensional analytic FFT generating function." *Bioinformatics* 24: 1865-1873.

[35] Ritchie DW, and Venkatraman V. 2010. "Ultra-Fast FFT Protein Docking On Graphics Processors." *Bioinformatics* 26: 2398-2405.

[36] Sebokova E, Christ AD, Boehringer M, and Mizrahi J. 2007. "Dipeptidyl peptidase IV inhibitors: the next generation of new promising therapies for the management of type 2 diabetes." *Curr Top Med Chem.* 7:547-555.

[37] Sicree R, Shaw J, and Zimmet P. 2006. "Diabetes and impaired glucose tolerance. Diabetes Atlas. International Diabetes Federation." 3rd ed. Belgium: *International Diabetes Federation* 15-103.

[38] Tabish SA. 2007. "Is Diabetes Becoming the Biggest Epidemic of the Twenty-first Century"? *Int. J. of Health Sci.* 1(2), V–VIII.

[39] Teilum K, Hoch JC, Goffin V, Kinet S, Martial JA, and Kragelund BB. 2005. Solution structure of human prolactin. *J. of Mol Bio.* 351:810–823.

[40] Uma AN, Sahithi P, and Rao PR. 2015. "*In-silico* Structure based, QSAR and Analogue based Studies using Dipeptidyl peptidase-4 (DPP4) Inhibitors against Diabetes Type-2." *Int. J. of Biotech and Biomed. Sci.* 1: 16-20.

[41] Wb AE. 2004. "Dipeptidyl peptidase IV inhibitors for the treatment of diabetes." *J. Med Chem.* 47: 4135-4141.

[42] Wiederstein M, and Sippl MJ. 2007. "ProSA-web: interactive web service for the recognition of errors in three-dimensional structures of proteins." *Nucleic Acids Research.* 35: 407-410.

[43] Zander M, Madsbad S, Madsen JL, and Holst JJ. 2000. "Effect of 6-week course of glucagon-like peptide 1 on glycaemic control, insulin sensitivity, and β-cell function in type 2 diabetes: a parallel-group study." *Lancet.* 359:824-830.

[44] Zhang Z, Schwartz S, Wagner L, and Miller W. 2000. "A greedy algorithm for aligning DNA sequences." *J. Comput Biol.* 7: 203-214.

[45] Zhaolan L, Chaowei F, Weibing W, and Biao X. 2010. "Prevalence of chronic complications of type 2 diabetes mellitus in outpatients-a correstional hospital based survey in urban China." *HQLO* 8: 62-71.

[46] Zimmet P, Alberti KG, and Shaw J. 2001. "Global and societal implications of the diabetes epidemic." *Nature* 414:782–7.

Chapter 4

HOMOLOGY MODELLING AND STRUCTURAL VALIDATION OF TYPE 2 DIABETES ASSOCIATED TRANSCRIPTION FACTOR 7-LIKE 2 (TCF7L2)

Rajneesh Prajapat[*], *Ijen Bhattacharya*
and Anoop Kumar
Department of Medical Biochemistry,
Faculty of Medical Sciences, Rama Medical College
and Hospital (NCR), Rama University, Kanpur, UP, India

ABSTRACT

New research findings indicate that variation in the transcription factor 7-like 2 (TCF7L2) gene, linked to the pathogenesis of type 2 diabetes. In the present study, the protein structure model of TCF7L2 was generated, to understand the structure, function and mechanism of the action of proteins. The stereo chemical quality of the protein model was checked by

---

[*] Corresponding Author's Email: prajapat.rajneesh@gmail.com.

using *in silico* analysis with PROCHECK, WHATIF, ProSA and QMEAN servers. The result of the study may be a guiding point for further investigations on TCF7L2 protein and its role in metabolic risk factors of type 2 diabetes. The 67.9% residue in the core region of Ramachandran plot showing high accuracy of protein model and the QMEAN Z-score of -6.07 indicates the overall model quality of TCF7L2 protein.

Keywords: homology modelling, TCF7L2, PROCHECK

INTRODUCTION

Type 2 diabetes is associated with impaired insulin secretion (1). The Transcription factor 7-like 2 (TCL7L2) gene product is a high mobility group (HMG) box-containing transcription factor implicated in blood glucose homeostasis (2). The TCF7L2 regulates genes involved in cell proliferation and differentiation. The TCF7L2 gene is located on chromosome 10q25 in a region of replicated linkage to type 2 diabetes (6). TCF7L2 has recently been implicated in the pathogenesis of type 2 diabetes (T2D) through regulation of pancreatic β-cell insulin secretion (7; 10).

The variants in TCF7L2 increase the risk for type 2 diabetes and novel evidence that the variants likely influence both insulin secretion and insulin sensitivity (18). Bioinformatics helps in management of complex biological data, sequence analysis and algorithmic designing (16). However, by using the *in silico* analysis we can analyze the protein sequences (12). Therefore, the present study enlists some of the physiochemical and functional properties of TCF7L2 protein and provides information about its three-dimensional structure.

4.1. METHODOLOGY

4.1.1. Operating System

The present study was conducted by using Intel (R) Core (TM) i3-370 M CPU @ 2.40 GHz and 32 bit operating system (HP ProBook).

4.1.2. Sequence Retrieval, Alignment and Homology Modelling

The FASTA sequence of transcription factor 7-like 2 (TCF7L2 [CAG38811]) protein was retrieve from NCBI. The PDB file of TCF7L2 [CAG38811] protein was generated by Phyre 2 servers by using its FASTA sequence. In order to build a model of protein domain, Multiple Sequence Alignment was performed between full length TCF7L2 protein sequence and another protein sequences in this database. To build the model of the TCF7L2 protein with more homology, structure of TCF7L2 protein model in 3D-JIGSAW server was selected as template. Model construction and regularization (including geometry optimization) of model were done by optimization protocol in YASARA. The energy of the model was minimized using the standard protocols of combined application of simulated annealing, conjugate gradient and steepest descent.

4.1.3. Model Reputation

The UCLA-DOE server provides a visual analysis of the quality of a putative crystal structure for protein. Verify 3D expects this crystal structure to be submitted in PDB format (9). The validation for structure models was performed by using PROCHECK (8). The accuracy of predicted model and its stereo chemical properties were evaluated by PROCHECK-NMR (19). The model was selected based on various factors such as overall G-factor, number of residues in core that fall in generously allowed and disallowed regions in Ramachandran plot (Figure 4.2). The model was further analyzed by WHATIF (8), QMEAN (9, 11) and ProSA (20). ProSA was used for the display of Z-score and energy plots.

4.2. Results

4.2.1. Building of Protein Model

Sequence alignment of TCF7L2 protein by using the phyre 2 server, revealed sequence homology with catenin binding domain (ID = 99%), which was selected as template for the model building of TCF7L2 protein. Total 41 residues (7% of query sequence) was modeled with 99% confidence by the single highest scoring template. To build the model, PSI-BLAST was done with the maximum E-value allowed for template being 0.005. Using catenin binding domain sequence Modelling of TCF7L2 protein domains was done with the help of YASARA (Figure 4.1).

4.2.2. Model Reputation

The model showed good stereo chemical property in terms of overall G-factor value of -0.64 indicating that geometry of the model corresponds to the probability conformation with 67.9% residues in the core region of Ramachandran plot showing high accuracy of model predicted. The number of residues in allowed and generously allowed region was 20.5% and 11.5% respectively, and none of the residues were present in the disallowed region of the plot (Figure 4.2).

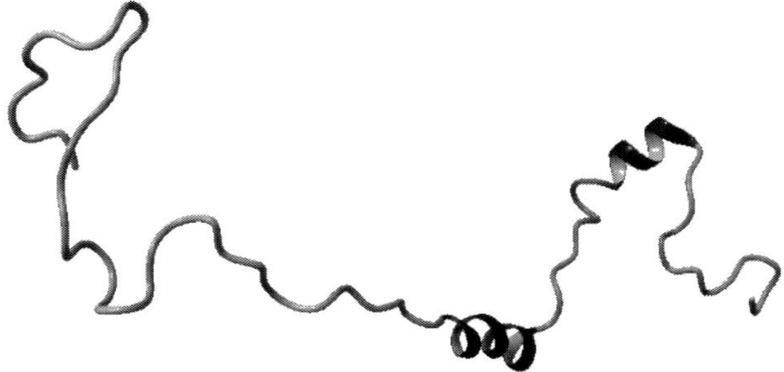

Figure 4.1. TCF7L2 protein ribbon model generated using YASARA.

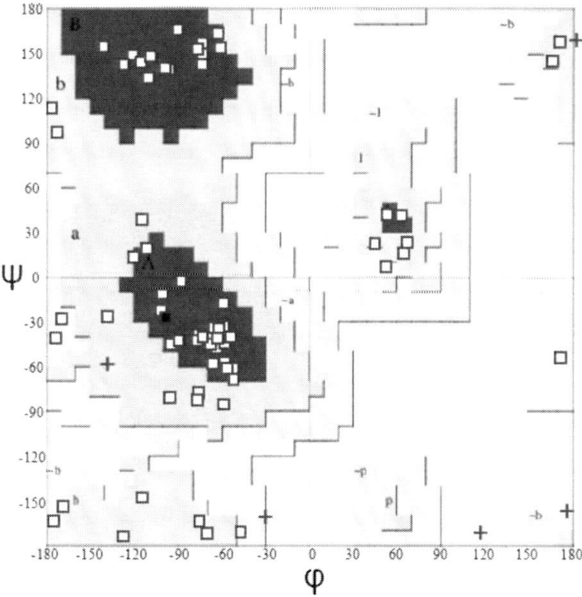

Figure 4.2. Ramachandran Plot analysis of TCF7L2 protein. Total number of residues were 156 with 67.9% in most favored regions [A, B, L], 20.5% in additional allowed regions [a,b,l,p], 11.5% in generously allowed regions and 0% in disallowed regions.

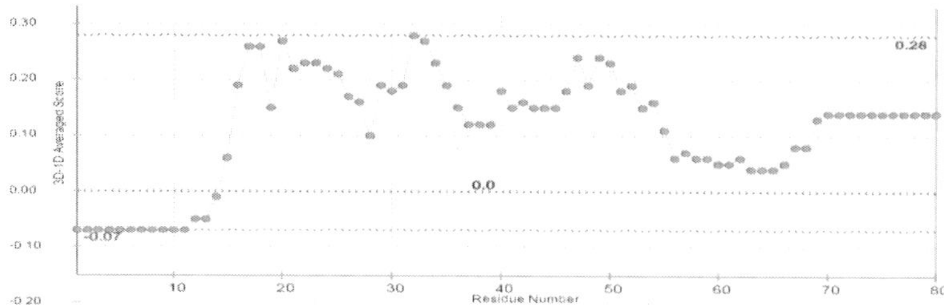

Figure 4.3. Verified 3D graph of TCF7L2 protein [CAG38811].

The high score of 0.28 indicates that environment profile of the model is good (Figure 4.3). The profile score above zero in Verify 3D graph (5; 9) corresponds to acceptable environment of the model. In Verified 3D plot, 17.50% of the residues had an averaged 3D-1D score >= 0.2.

4.2.3. Model Validation

ProSA was used to check the three-dimensional model of TCF7L2 proteins for potential errors. The ProSA Z-score of -6.07 indicates the overall model quality of TCF7L2 protein (Figure 4.4).

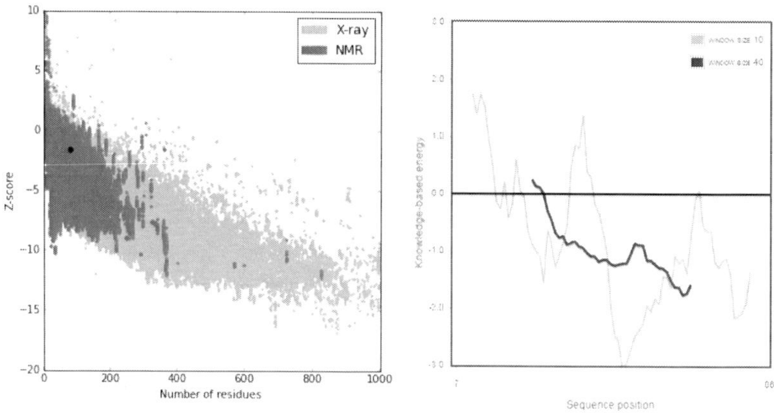

Figure 4.4. ProSA web service analysis of TCF7L2 protein model.

The QMEAN score of the model was 0.189 and the Z-score was -4.16, which was very close to the value of 0 and this shows the fine quality of the model because the estimated reliability of the model was expected to be in between 0 and 1 and this could be inferred from the density plot for QMEAN scores of the reference set (Fig. 4.5A). A comparison between normalized QMEAN score (0.40) and protein size in non-redundant set of PDB structures in the plot revealed different set of Z-values for different parameters such as C-beta interactions (-1.22), interactions between all atoms (-1.46), solvation (-0.39), torsion (-3.83), SSE agreement (-0.53) and ACC agreement (-2.88) (Figure 4.5B).

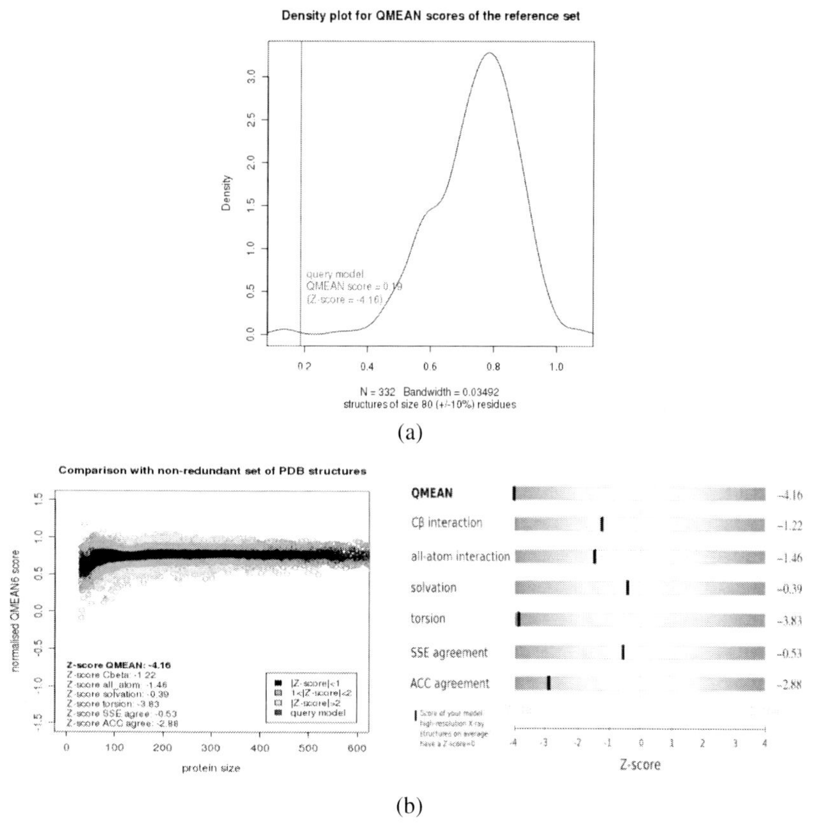

(a)

(b)

Figure 4.5. (a) The density plot for QMEAN showing the value of Z-score and QMEAN score (b) Plot showing the QMEAN value as well as Z-score.

CONCLUSION

The generated model could be supportive to understand the functional characteristics of transcription factor 7-like 2 (TCF7L2). The variants in TCF7L2 associated with the risk for type 2 diabetes (17). The *in silico* molecular Modelling and validation studies is helpful to understand the structure, function and mechanism of proteins action (14). The structure validation of generated model was done by using WHATIF, PROCHECK, ProSA and QMEAN confirmed the reliability of the model.

The model showed good stereo-chemical property in terms of overall G-factor value of -0.64 indicating that geometry of model corresponds to the probability conformation with 67.9% residue in the core region of Ramachandran plot showing high accuracy of model prediction (13). The Z score of -6.07 predicted by ProSA represents the good quality of the model. Z-score also measures the divergence of total energy of the structure with respect to an energy distribution derived from random conformations. The scores indicate a highly reliable structure and are well within the range of scores typically found for proteins of similar size (21, 22). The energy plot shows the local model quality by plotting knowledge-based energies as a function of amino acid sequence position (15). QMEAN analysis was also used to evaluate and validate the model.

REFERENCES

[1]   Anna LG, Braun M, and Rorsman P. 2009. "Type 2 Diabetes Susceptibility Gene TCF7L2 and Its Role in β-Cell Function." *Diabetes* 58: 800–802.

[2]   Agrawal P, Thakur Z, and Kulharia M. 2013. "Homology modelling and structural validation of issue factor pathway inhibitor." *Bioinformation* 9: 808-812.

[3] Benkert P, Künzli M, and Schwede T. 2009. "QMEAN server for protein model quality estimation." *Nucleic Acids Res.* 37: 4-10.

[4] Benkert P, Tosatto SC, and Schomburg D. 2008. "QMEAN: A comprehensive scoring function for model quality assessment." *Proteins* 71: 261-267.

[5] Bowie JU, Luthy R, and Eisenberg D. 1991. "A method to identify protein sequences that fold into a known three-dimensional structure." *Science* 253:164-170.

[6] Damcott CM, Pollin TI, Reinhart LJ, Ott SH, Shen H, Silver KD, Mitchell BD, and Shuldiner AR. 2006. "Polymorphisms in the transcription factor 7-like 2 (TCF7L2) gene are associated with type 2 diabetes in the Amish: replication and evidence for a role in both insulin secretion and insulin resistance." *Diabetes* 55:2654-9.

[7] Grant SF, Thorleifsson G, Reynisdottir I, Benediktsson R, Manolescu A, Sainz J, Helgason A, Stefansson H, Emilsson V, Helgadottir A, Styrkarsdottir U, Magnusson KR, Walters GB, Palsdottir E, Jonsdottir T, Gudmundsdottir T, Gylfason A, Saemundsdottir J, Wilensky RL, Reilly MP, Rader DJ, Bagger Y, Christiansen C, Gudnason V, Sigurdsson G, Thorsteinsdottir U, Gulcher JR, Kong A, and Stefansson K. 2006. "Variant of transcription factor 7-like 2 (TCF7L2) gene confers risk of type 2 diabetes." *Nat. Genet.* 38:320–323.

[8] Laskowski RA, MacArthur MW, Moss DS, and Thornton JM. 1993. "PROCHECK: a program to check the stereochemical quality of protein structures." *J. Appl. Cryst.* 26:283–291.

[9] Luthy R, Bowie JU, and Eisenberg D. 1992. "Assessment of protein models with three-dimensional profiles." *Nature* 356: 83-85.

[10] Ng MC, Tam CH, Lam VK, So WY, Ma RC, and Chan JC. 2007. "Replication and Identification of Novel Variants at TCF7L2 Associated with Type 2 Diabetes in Hong Kong Chinese." *J. Clin. Endocrinol Metab.*, 92:3733-7.

[11] Novotny WF, Girard TJ, Miletich JP, and Broze GJJJ. 1988. "Platelets secrete a coagulation inhibitor functionally and

antigenically similar to the lipoprotein associated coagulation inhibitor." *Blood* 72: 2020-2025.

[12] Pevzner P, and Shamir R. 2011. *Bioinformatics for Biologists*: Cambridge University Press."

[13] Prajapat, R, Marwal A, Bajpai V, and Gaur RK. 2011. "Genomics and Proteomics Characterization of Alphasatellite in Weed Associated with Begomovirus." *Int. J. of Plant Pathology* 2: 1-14.

[14] Prajapat R, Avinash M, Shaikh Z, and Gaur RK. 2012. "Geminivirus Database (GVDB): First Database of Family Geminiviridae and its genera Begomovirus." *Pakistan J. of Biological Sci.* 15: 702-706.

[15] Prajapat R, Marwal A, Gaur RK. 2014. "Recognition of Errors in the Refinement and validation of three-dimensional structures of AC1 proteins of begomovirus strains by using ProSA-Web." *Journal of Viruses.* doi.org/10.1155/2014/752656.

[16] Rasouli H, and Fazeli-Nasab B. 2014. "Structural Validation and Homology modelling of Lea 2 Protein in Bread Wheat." *American-Eurasian J. Agric. & Environ. Sci.* 14: 1044-1048.

[17] Shah M, Ron TV, John MM, Piccinini F, Man CD, Cobelli C, Kent RB, Rizza RA, and Adrian V. 2015. "TCF7L2 genotype and α-cell function in nondiabetic humans." *Diabetes* doi: 10.2337/db15-1233.

[18] Shu L, Sauter NS, Schulthess FT, Matveyenko AV, Oberholzer J, and Maedler K. 2008. "Transcription factor 7-like 2 regulates beta-cell survival and function in human pancreatic islets." *Diabetes* 57:645-53.

[19] Vriend G. 1990. "WHAT IF: a molecular Modelling and drug design program." *Journal of Molecular Graph.* 8: 52-56.

[20] Wiederstein M, and Sippl MJ. 2007. "ProSA-web: interactive web service for the recognition of errors in three-dimensional structures of proteins." *Nucleic Acids Research.* 35: 407-410.

[21] Wiederstein M, and Sippl MJ. 2005. "Protein sequence randomization: efficient estimation of protein stability using knowledge-based potentials." *J. Mol. Biol.* 345: 1199–1212.

[22] Yi F, Brubaker PL, and Jin T. 2005. "TCF-4 mediates cell type-specific regulation of proglucagon gene expression by beta-catenin and glycogen synthase kinase-3-beta." *J. Biol. Chem.* 280: 1457-1464.

In: Medical Bioinformatics and Biochemistry   ISBN: 978-1-53614-952-4
Editors: Rajneesh Prajapat et al.   © 2019 Nova Science Publishers, Inc.

Chapter 5

EFFECT OF VITAMIN E AND C SUPPLEMENTATION ON OXIDATIVE STRESS IN DIABETIC PATIENTS

Rajneesh Prajapat**[*] and **Ijen Bhattacharya
Department of Medical Biochemistry,
Faculty of Medical Sciences, Rama Medical College
and Hospital (NCR), Rama University, Kanpur, UP, India

ABSTRACT

Diabetes is a metabolic epidemic that causes vascular complications. The presence of oxidative stress in diabetes patients and the preventive role of vitamins therapy have been reported by many researchers. Vitamins supplementation improves antioxidant defense system in diabetes patients. Subjects enrolled in the study received 500 mg of both vitamin C and vitamin E daily twice for 4 months under medical supervision. Fasting blood glucose, MDA, catalase, SOD and nitric oxide were determined. Serum vitamin E and plasma vitamin C were also measured. Statistical

---

[*] Corresponding Author's Email: prajapat.rajneesh@gmail.com.

analysis was performed using SPSS 12.0. Numerical normally distributed and categorical data were compared using independent t-test. Data were presented as means ± SD. After supplementation with vitamin E and C in diabetic patients, a signify decrease in FBS, MDA levels and increase in serum nitrite, erythrocyte SOD, blood catalase, vitamin E and vitamin C levels were observed. A negative correlation between MDA and vitamins was observed. Vitamin E and C supplementation is useful for the treatment of oxidative stress related complications in diabetes patients. Prescribed medicines contain active ingredients that may causes effect on the patients in terms of side effects. Controlled vitamin therapy for prolonged period not causes any side effects, as well as play effective role for the management of type 2 diabetic related oxidative stress.

Keywords: type 2 diabetes, oxidative stress, vitamin E and C

INTRODUCTION

Diabetes mellitus is metabolic epidemic that occur due to ineffective secretion of insulin from pancreas [1, 2] and the possibly the worldwide number of diabetic patient will be rise to 439 million by 2030 [3, 4]. Vitamin E supplementation could improve glycemic control in diabetes patients [5, 6]. Vitamin C supplementation is effective in prevention of non-enzymatic glycosylation of proteins [7] and it also serves as therapeutic agent for diseases that protects body from damage caused by free radicals [8].

The reactive oxygen species (ROS) can damage cellular bio-molecules (protein, nucleic acid etc.) as well as plasma membrane. The malondialdehyde (MDA) is reported as indicator of lipid peroxidation reactions [9]. Involvement of Nitric oxide (NO•) in smooth muscle relaxation, cytotoxic reactions, and neuronal transmission [10] is reported in the literature. Currently researchers are focusing on the role of NO• in the development of uremic symptoms [11, 12].

In addition, some vitamins can prevent the harmful effects of free radicals by non-enzymatic modes like in both vitamin E (α-tocopherol) and C (ascorbic acid). The α-tocopherol prevents damage to polyunsaturated fatty acids by free radicals in membranes [13].

The present study planned to investigate the possible alterations of oxidant - antioxidant status in diabetic patients and effect of vitamin E and C supplementation. This study included fifty diabetic patients and control, in the mean age of 52.3 ± 9.62 years.

5.1. METHODOLOGY

Present study was performed at the Rama Hospital, Rama Medical College (NCR), UP [India], as a randomized controlled trial and with parallel design. According to ADA [14], the fifty patients with type 2 diabetic of mean age 52.3 ± 9.62 years were selected for the study. Alcoholics, Smokers, patients with chronic inflammatory conditions, or hepatic or respiratory diseases were excluded from the study.

The study was reviewed and approved by Ethics Committee, Rama hospital, Rama Medical College (NCR), UP [India]. Subjects enrolled in the study received 500 mg of vitamin C and 500 mg of vitamin E, daily twice for 4 months under medical supervision. The age, weight, sex, height, diabetes duration, blood pressure was examined ($\geq 130/\leq 80$ mm Hg) and recorded. Venous blood (10 ml) was collected by using standard laboratory methods at each study point and used for various estimations.

Malondialdehyde (MDA) was estimated by the modified protocol of Mossa et al. [15]. MDA in serum was separated and determined as conjugate with TBA. The MDA-TBA complex was measured at 534 nm. Catalase activity was determined by modified protocol of Goth [16]. Serum catalase activity is linear up to 100 kU/l.

The SOD activity was assayed by the modified method of Kakkar et al. [17]. The SOD activity was measured by the inhibition of the reduction of nitroblue-tetrazolium by Superoxide anion produced by potassium Superoxide (K^+, O_2^-) dissolved in dimethyl sulfoxide. Nitric oxide was determined by Cortas and Wakid method [18]. Nitric oxide is a labile and diffusible molecule, which forms stable metabolites (nitrite/nitrate, NO_2^- and NO_3^-), which are detected by Griess reaction.

Serum vitamin E was measured by their reduction of ferric to ferrous ion, which then forms a red colored complex with α-α'-bipyridylas in Baker and Frank method [19]. Plasma vitamin C was determined by DNPH method [20], where vitamin C is oxidized to diketogulonic acid, which reacts with 2,4 dinitrophenylhydrazineto form diphenylhydrozone.

Statistical analysis was performed using Statistical Package for Social Sciences (SPSS 12.0, Chicago IL). Numerical normally distributed data and categorical data were compared using independent t-test. Results were given with their 95% CIs. Data were presented as means ± SD. Numerical normally distributed data and categorical data were compared using independent t-test.

5.2. RESULTS AND DISCUSSION

All the selected type 2 Diabetes mellitus patients were supplemented with 500 mg/day twice of both vitamin E and C for four months, after two months of supplementation all the parameters mentioned above were studied again.

Reduced vitamin E and C levels were lower and malondialdehyde levels (MDA) were higher in type 2 diabetes mellitus patients compared to healthy controls ($p < 0.001$). Table depicts, after vitamin supplementation, causes significant reduction in fasting blood sugar [FBS], significant increase ($p < 0.001$) in concentration of SOD, Nitrite, catalase in type 2 DM group as compared to controls. Significant fall ($p < 0.001$) in antioxidants like serum MDA was observed in type 2 DM as compared controls. P* < 0.05 **<0.001 compared to before treatment [Table 5.1, Figure 5.1].

Table 5.1. Levels of biochemical parameters in diabetic patients and controls before and after supplementation with doses of vitamin C and E [(Data are mean ± SD)

Parameters (Tests)	Control	Diabetic Group [Mean ± SD]	Diabetic with Vitamin E and C Supplementation		P Value
			2 Month [Mean ± SD]	4 Months [Mean ± SD]	
FBS (mg/dl)	102 ± 4.64	162.78 ± 4.82	147.09 ± 7.05	49.73	0.00001
Serum MDA (μmol/l)	1.95 ± 0.54	3.72 ± 0.30	3.00 ± 0.41	2.74 ± 0.40	< .001
Erythrocyte SOD (U/ mg Hb)	1.54 ± 0.62	0.82 ± 0.10	1.02 ± 0.18	1.20 ± 0.38	< .001
Serum Nitrite (μmol/l)	67.42 ± 4.08	59.26 ± 7.4	64.52 ± 4.5	66.94 ± 5.5	< .005
Blood Catalase (k/gm Hb)	291.55 ± 13.47	210.08 ± 23.78	233.62 ± 19.54	244.58 ± 21.81	< .001
Serum Vitamin E (mg/dl)	1.24 ± 0.21	0.86 ± 0.06	1.75 ± 0.21	1.92 ± 0.22	< .001
Plasma Vitamin C (mg/dl)	1.55 ± 0.41	0.78 ± 0.08	1.32 ± 0.21	1.47 ± 0.16	< .001

Values were expressed as Mean ± SD, indicates p<0.001 (unpaired 't' test).

A negative correlation of MDA with vitamin E and vitamin C was found. Figure 1 describes about standard deviation showing mean values and positive, negative error of serum MDA, erythrocyte SOD, serum nitrite, blood catalase, vitamin E and vitamin C before and after supplementation/treatment with doses of vitamin C and E [(Data are mean ± SD)].

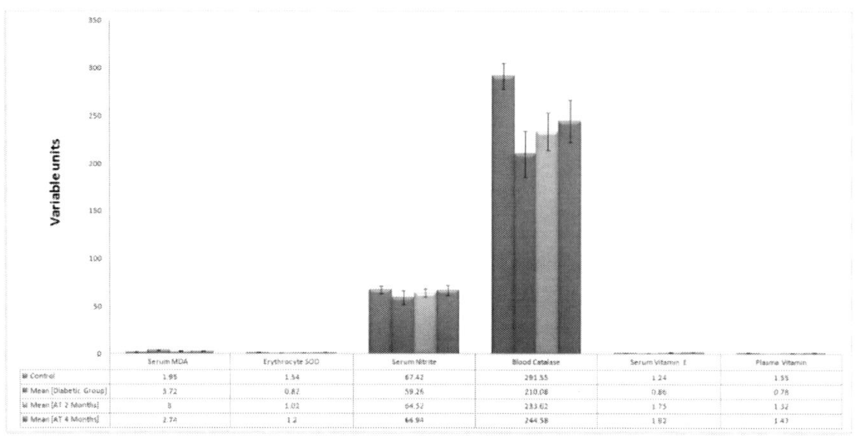

	Serum MDA	Erythrocyte SOD	Serum Nitrite	Blood Catalase	Serum Vitamin E	Plasma Vitamin
Control	1.95	1.54	67.42	291.55	1.24	1.55
Mean [Diabetic Group]	3.72	0.82	59.26	210.08	0.86	0.78
Mean [AT 2 Months]	3	1.02	64.52	233.62	1.75	1.32
Mean [AT 4 Months]	2.74	1.2	66.94	244.58	1.82	1.42

Figure 5.1. Standard deviation graph showing mean values and positive, negative error of serum MDA, erythrocyte SOD, serum nitrite, blood catalase, vitamin E and vitamin C of control group [blue bars], diabetic group before [red bars], after two months [green bars], and after four month [purple bars] of supplementation/treatment with doses of vitamin C and E [(Data are mean ± SD)].

Vitamin E ameliorates oxidative stress in type 2 diabetes mellitus patients and improves antioxidant defense system. However, vitamin E does not have any advantage for metabolic parameters [21]. In the present study, the increased serum MDA level in diabetic patients indicates that indeed there is oxidative stress. Further increase in serum MDA levels shows that the oxidative stress has increased in these patients. Oxidative stress may increase the synthesis of asymmetric dimethyl arginine (ADMA), which is an endogenous inhibitor of endothelial nitric oxide synthase [22].

The most probable explanation for decreased SOD activity is a possible direct inactivation of the enzyme by its product hydrogen peroxide, or by superoxide anion itself [23]. Decreased SOD activity could also be related to trace element deficiencies in patients [24].

Significant decrease in the activity of catalase could be due to less availability of NADPH [25]. Our study shows a significant decrease in catalase activity. This decrease could be due to increase in MDA [26]. Vitamin E is a lipophilic antioxidant. Vitamin E radical formed by free radical attack interact with vitamin C and regenerate vitamin E. In the process vitamin C is consumed and vitamin E is formed [27, 28].

CONCLUSION

Previous research findings explain the presence of oxidative stress in diabetes patients and the preventive role of vitamins therapy. Results of present study suggested that, vitamin E and C supplementation is useful for the treatment of oxidative stress related complications in diabetes patients. Prescribed medicines contain active ingredients that may causes effect on the patients in terms of side effects. Controlled vitamin therapy for prolonged period not causes any side effects, as well as play effective role for the management of type 2 diabetic related oxidative stress.

REFERENCES

[1]   Pandit A, and Pandey AK. 2016. "Estimated Glomerular Filtration Rate and Associated Clinical and Biochemical Characteristics in Type 2 Diabetes Patients." *Adv in Diabetes and Metabol.* 4: 65 - 72. doi: 10.13189/adm.2016.040402.

[2]   Awasthi A, Parween N, Singh VK, Anwar A, Prasad B, and Kumar J. 2016. "Diabetes: Symptoms, Cause and Potential Natural Therapeutic Methods." *Adv in Diabetes and Metabol.* 4: 10-23. doi: 10.13189/adm.2016.040102.

[3]   Prajapat R, and Bhattacharya I. 2016. "*In-silico* Structure Modelling and Docking Studies Using Dipeptidyl peptidase-4 (DPP4) Inhibitors against Diabetes Type-2." *Adv in Diabetes and Metabol.* 4: 73-84. doi: 10.13189/adm.2016.040403.

[4]   Rahman Hassan SAEL, Rahman Elsheikh WA, Abdel Rahman NI, and Nabiela M. ElBagir. 2016. "Serum Calcium Levels in

Correlation with Glycated Hemoglobin in Type 2 Diabetic Sudanese Patients." *Adv in Diabetes and Metabol.* 4: 59 - 64. doi: 10.13189/adm.2016.040401.

[5]   Prajapat R, Bhattacharya I, and Jakhalia A. 2017. "Combined Effect of Vitamin C and E Dose on Type 2 Diabetes Patients." *Adv in Diabetes and Metabol.* 5: 21-25. doi: 10.13189/ adm.2017.050201.

[6]   Khabaz M, Rashidi M, Kaseb F, and Afkhami-Ardekan M. 2009. "Effect of Vitamin E on Blood Glucose, Lipid Profile and Blood Pressure in Type 2 Diabetic Patients." *Iranian J of diabetes and Obesity* 1: 10-15.

[7]   Dakhale GN, Chaudhari HV, and Shrivastava M. 2011. "Supplementation of Vitamin C Reduces Blood Glucose and Improves Glycosylated Hemoglobin in Type 2 Diabetes Mellitus: A Randomized, Double-Blind Study." *Adv in Pharmacological Sciences*, vol. 2011. doi:10.1155/2011/195271.

[8]   Chambial S, Dwivedi S, Shukla KK, John PJ, and Sharma P."Vitamin C in Disease Prevention and Cure: An Overview." *Indian J Clin Biochem.* 28: 314–328. doi: 10.1007/s12291-013-0375-3.

[9]   Sharma BK, Sagar S, Kallo IJ, Karl N, and Ganguly NK. 1992. "Oxygen free radical in essential hypertension." *Molecular and Cellular Biochem.* 111: 103-108.

[10] Amero A, Bonaudo R, Ghigo D, Arese M, Costamagna C, and Cirina P et al., "Enhanced production of nitric oxide by blood - dialysis membrane interaction." *J Am Soc Nephrol.* 6:1278-1283.

[11] Sarkar SR, Kaitwatcharachai C, and Levin NW. 2004. "Nitric oxide and hemodialysis." *Semin Dial.* 17: 224-228.

[12] Lin SH, Chu P, Yu FC, Diang LK, and Lin YF. 1996. "Increased nitric production in hypotensive hemodialysis patients." *ASAIO J* 42: M895-9.

[13] Aymelek G, Yesim A, Mehmat NO, and Bolkan S. "Lipid peroxidation and antioxidant systems in hemodialyzed patients." *Dial Transplant* 31:88-96.

[14] American Diabetes Association. Standards of medical care in diabetes-2006. *Diabetes Care* 29:4-42.

[15] Mossa MM, Bushra MM, Salih MR, and May NY. 2009. "Estimation of malondialdehyde as oxidative factor and glutathione as early detectors of hypertensive pregnant women." *Tikrit Medical Journal* 15(2): 63-69.

[16] Goth L. 1991. "A simple method for determination of serum catalase activity and reversion of reference range." *Clinics Chimica Acta* 196:143-152.

[17] Kakkar P, Das B, and Viswanathan PN. 1984. "A modified spectrophotometric assay of superoxide dismutase." *Indian J Biochem Biophys.* 21:130-132.

[18] Cortas NK, and Wakid NW. 1990. "Determination of inorganic nitrate in serum and urine by a kinetic cadmium reduction method." *Clin Chem.* 36:1440- 3.

[19] Baker, and Frank. "Determination of serum tocopherol by colorimetric method." 1988. In: Gowenlock AH ed. *Varley's Practical Clinical Biochemistry*, Heinemann Professional publishing pp.902-3.

[20] Castelli A, Martorana GE, Frasca AM, and Meucci E. 1981. "Colorimetric determination of plasma vitamin C: comparison between 2,4-dinitrophenylhydrazine and phosphotungstic acid methods." *Acta Vitaminol Enzymol* 3:103-10.

[21] Gupta S, Sharma TK, Kaushik GG, and Shekhawat VP. 2011. "Vitamin E supplementation may ameliorate oxidative stress in type 1 diabetes mellitus patients." *Clin Lab.* 57:379-86.

[22] Saran R, Novak JE, Desai A, Abdhulhayoglu E, Warren JS, and Bustami R et al., 2003."Impact of vitamin E on plasma asymmetric dimethylarginine (ADMA) in chronic kidney disease (CKD): a pilot study." *Nephrol Dial Transplant* 18: 2415-20.

[23] Salo DC, Lin SW, Pacifici RE, and Davies KJ. 1998. "Superoxide dismutase is preferentially degraded by a proteolytic system from red blood cells following oxidative modification by hydrogen peroxide." *Free Radic Biol Med.* 5: 335-39.

[24] Sinet PM, and Garber P. 1981. "Inactivation of the human Cu-Zn superoxide dismutase during exposure to O_2 and H_2O_2." *Arch Biochem Biophys.* 212: 411-6.

[25] Hernandez de Rojas A, and Martin Mateo MC. 1996. "Superoxide dismutase and catalase activities in patients undergoing hemodialysis and continuous ambulatory peritoneal dialysis." *Renal Failure.* 18: 937-46.

[26] Chauhan DP, Gupta PH, Namporthic MRN, and Singal PC. 1982. "Determination of RBC superoxide dismutase, catalase, G- 6-PD, reduced glutathione and MDA in uremia." *Clin Chem Acta.* 123: 153-9.

[27] Bhogade RB, Suryakar AN, Joshi NG, and Patil RY. 2008. "Effect of Vitamin E Supplementation on Oxidative Stress in Hemodialysis Patients." *Indian J of Clinical Biochem.* 23:233-237.

[28] Aymelek G, Yesim A, Mehmat NO, and Bolkan S. 2002. "Lipid peroxidation and antioxidant systems in hemodialyzed patients." *Dial Transplant* 31: 88-96.

In: Medical Bioinformatics and Biochemistry   ISBN: 978-1-53614-952-4
Editors: Rajneesh Prajapat et al.   © 2019 Nova Science Publishers, Inc.

Chapter 6

IN SILICO STRUCTURE ANALYSIS OF TYPE 2 DIABETES ASSOCIATED CYSTEINE PROTEASE CALPAIN-10 (CAPN10)

Rajneesh Prajapat[*] and Ijen Bhattacharya
Department of Medical Biochemistry,
Faculty of Medical Sciences, Rama Medical College
and Hospital (NCR), Rama University, Kanpur, UP, India

ABSTRACT

The intracellular calcium (Ca^{2+}) level activates cysteine protease calpain10 (CAPN10). The calpain10 is known to be involved in disease such as cancer, stroke and heart attack. The role of cysteine protease calpain10 (CAPN10) was recently identified and associated with diabetes mellitus type 2. In this paper, homology modelling procedure was used to determine the 3D structure of human calpain10 (AAH07553). The μ-calpain (1QXP) of *Rattus norvegicus* was selected as template for the construction of calpain10 model. Ramachandran plot of calpain10

[*] Corresponding Author's Email: prajapat.rajneesh@gmail.com.

(AAH07553) has only 55.9% residues in the most favored region while template μ-calpain (1QXP) has 69.3% residues in the most favored region. The model was validated by using protein structure tools RAMPAGE and Prochek for reliability. 3D structure of calpain10 suggested its active site remain conserved among family members and the major interactions are like those observed for template (1QXP).

Keywords: calpain10, Ramachandran plot, 1QXP

INTRODUCTION

The type 2 diabetes (T2D) is triggered by environmental and genetic risk factors and characterized by impaired insulin stimulated glucose uptake in muscle; improved hepatic glucose production and altered glucose induced insulin secretion [10, 33]. The CAPN10 gene encodes calpain10. Approximate 14 calpains (calcium-dependent cysteine proteases) isoforms are characterized that present in multiple tissues [29, 28, 23] those involved in diabetes, such as pancreas, liver and skeletal muscle [29]. CAPN10 abundant in the mitochondria and it is involved in apoptosis and age-related diseases, plus release and storage of Ca^{2+} [22]. The calpains are non-lysosomal cysteine proteases [26] that are activated by intracellular Ca^{2+}. An increase in the concentration of mitochondrial Ca^{2+} can initiate a series of proteolytic signals and participate in various signal transduction pathways [11] that can cause irreversible damage to cells. Therefore, the over expression of this enzyme is responsible for mitochondrial dysfunction [2].

The fine mapping of diabetes related genes suggested that the calpain10 (CAPN10) gene may serve as an important T2D susceptibility gene [31, 7]. A variation in the gene encoding the cysteine protease calpain10 (CAPN10) was recently linked and associated with type 2 diabetes [27, 25].

The biological effects of calpain proteases on phenotypes related to glucose homeostasis, a number provide support for the association of polymorphisms with type 2 diabetes [13]. The genetic variation at *CAPN10* in different human populations over a range of phenotypes related to type 2 diabetes, physiological studies on the biological functions of calpain proteases, and evolutionary studies on *CAPN10*. The CAPN10 influences insulin sensitivity and glucose homeostasis in nondiabetic members of kindreds at high risk for T2DM [9].

Homology modelling refers to modelling of protein 3D structure using a known experimentally determined structure of a homologous protein as a template. Homology modelling provides structural information important to understand of protein function, dynamics, interactions with ligands and other proteins [3]. In the present study, we used protein homology modelling [15] to determine the structure of human calpain 10.

6.1. METHODOLOGY

6.1.1. Operating System

In the present study Intel (R) Core (TM) i3-370 M CPU @ 2.40 GHz and 32-bit operating system (HP ProBook) was used.

6.1.2. Retrieval of Sequences and Sequence Alignment

The amino acid sequence of human calpain10 (AAH07553) was retrieved from GenBank-NCBI (www.ncbi.nlm.nih.gov) in the FASTA format. The calpain10 sequence was identified as one of the 46 members of the C2 family of proteases according to the MEROPS database [26]. To build a model of protein domain, Multiple Sequence Alignment was performed between full length calpain10 (AAH07553) sequence and aligned in the PDB server via the BLASTp alignment tool [1, 34] to search against the PDB (Protein Databank) to find out the related homologues. The PDB file of calpain10 (AAH07553) was generated by using 3D-JIGSAW protein comparative modelling servers. The PDB file of query and homologous template sequence were further utilized for 3D model energy validation [12].

6.1.3. Molecular Modelling of Calpain10

The UCLA-DOE server provides a visual analysis of the quality of a putative crystal structure for protein. Verify 3D expects this crystal structure to be submitted in PDB format [17]. RAMPAGE program was used for visualizing and assessing the Ramachandran plot of calpain10 (AAH07553) and its homologous template [16]. The validation for structure models was performed by using PROCHECK [14]. The model was selected on the basis of various factors such as overall G-factor, number of residues in core that fall in generously allowed and disallowed regions in Ramachandran plot (Figure 1; Figure 2). The model was further analyzed by QMEAN [4, 21] and ProSA [32]. ProSA was used for the display of Z-score and energy plots.

The 3D structure selected as a template for constructing the model of calpain10 was μ-calpain from *Rattus norvegicus* (PDB access code: 1QXP), which has a length of 900 amino acid residues and a resolution of 2.8 Å by X-ray crystallography [8]. It has 31% identity and 46% similarity with the target sequence (1QXP). The quality of folding was checked using Verify3D [17], and the system energy and quality of 3D alignment were monitored using Atomic Non-Local Environment Assessment (ANOLEA) [18, 19]. Produced models of query and homologous template were ranked on QMEAN server and utilized for the ribbon structural model construction (Figure 9, Figure 10). QMEAN locate the position of different amino acids present in the active site of proteins and estimates per-residue error [4].

6.2. RESULTS AND DISCUSSION

6.2.1. Building of Protein Model

Sequence alignment of calpain10 (AAH07553) protein by using the BLAST, revealed sequence homology with μ-calpain (1QXP) (ID= 31%), which was selected as template for the model building of calpain10 (AAH07553) protein. Total 227 residues (45% of query sequence) have been modelled with 99% confidence by the single highest scoring template. To build the model, BLAST was done with the maximum E-value allowed for template being $5e^{-62}$.

6.2.2. Model Reputation

The calpain10 (AAH07553) model showed good stereo chemical property in terms of overall G-factor value of -0.65 indicating that geometry of the model corresponds to the probability conformation with 55.9% residues in the core region of Ramachandran plot showing high accuracy of model predicted [24]. The number of residues in allowed and generously allowed region was 26.5% and 17.6% respectively, and none of the residues were present in the disallowed region of the plot (Figure 6.1). A similar approach was also used for template μ-calpain (1QXP) and its Ramachandran plot has 69.3% residues in favored region, 20.1% in the allowed region and 10.6% in the outlier regions (Figure 6.2).

The Ramachandran plot of calpain10 (AAH07553) has only 55.9% residues in the most favored region and homologous template μ-calpain (1QXP) has 69.3% residues in most favored regions therefore μ-calpain (1QXP) is more stable than calpain10 (AAH07553) (Table 6.1).

                    Rajneesh Prajapat and Ijen Bhattacharya

A good quality Ramachandran plot would be expected to have over 90% in the most favored regions [20]. Therefore, the energy minimization of both models should be required to enhance the stability by using the standard protocols of combined application of simulated annealing, conjugate gradient and steepest descent.

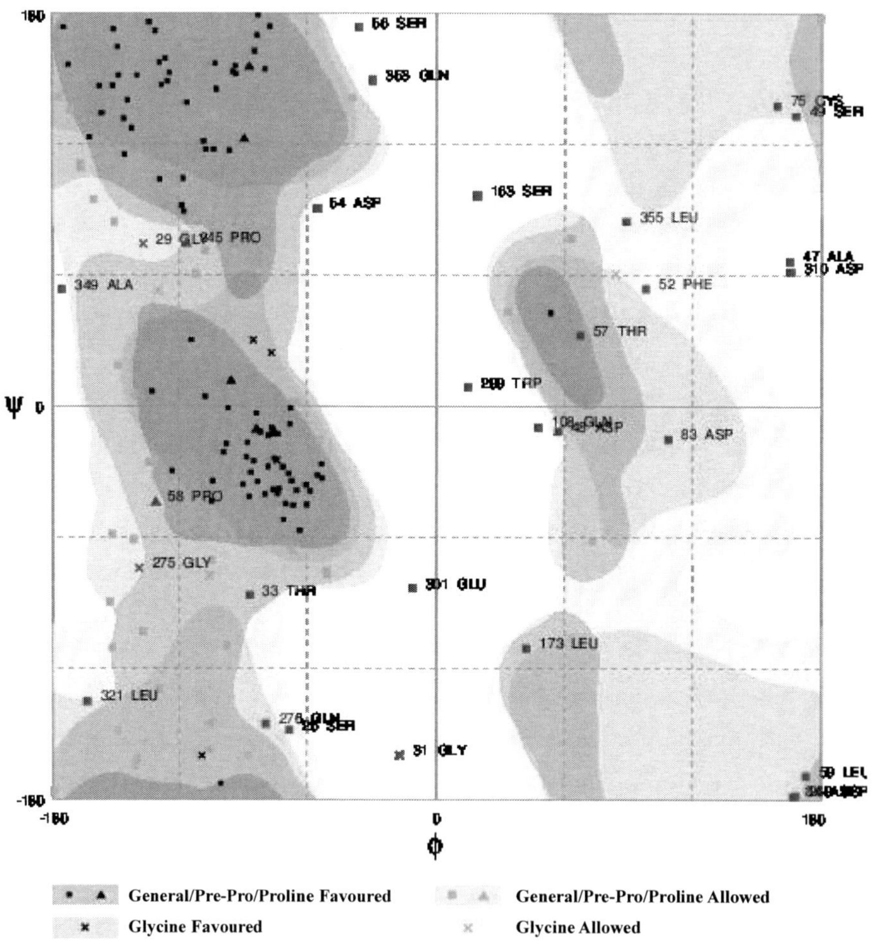

Figure 6.1a. Ramachandran plot of 3D model of calpain10 (AAH07553).

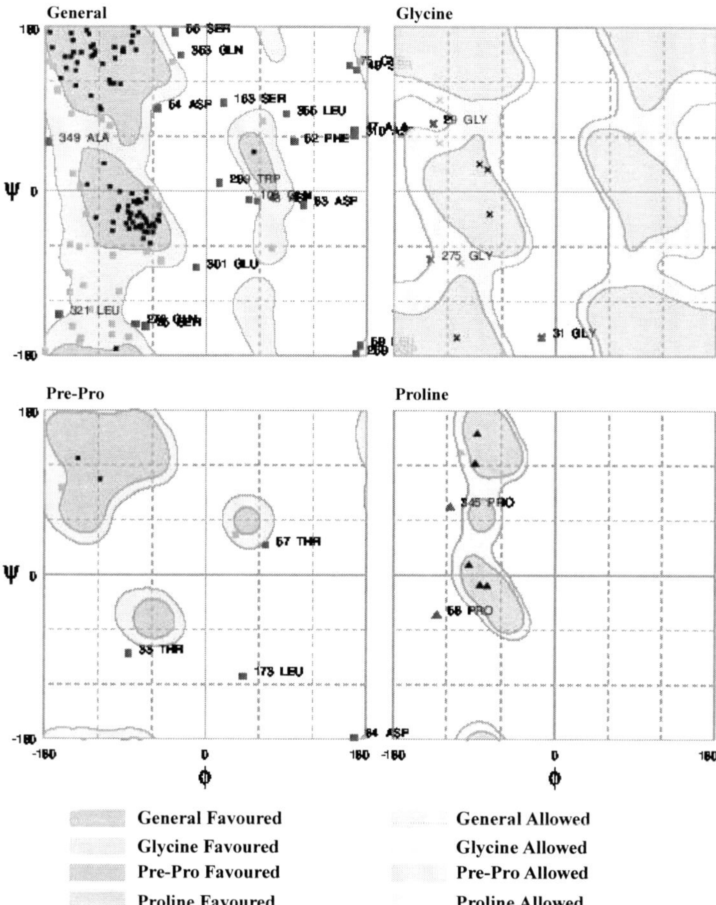

Figure 6.1b. Non-proline residues and non-glycine residue regions.

Table 6.1. Results summary of the Ramachandran plot

Accession No	Protein	Description	Residues (%)		
			Favored regions	Allowed regions	Outlier regions
AAH07553	calpain10	*Homo sapiens*	55.9	26.5	17.6
1QXP	μ-calpain	*Rattus norvegicus*	69.3	20.1	10.6

          Rajneesh Prajapat and Ijen Bhattacharya

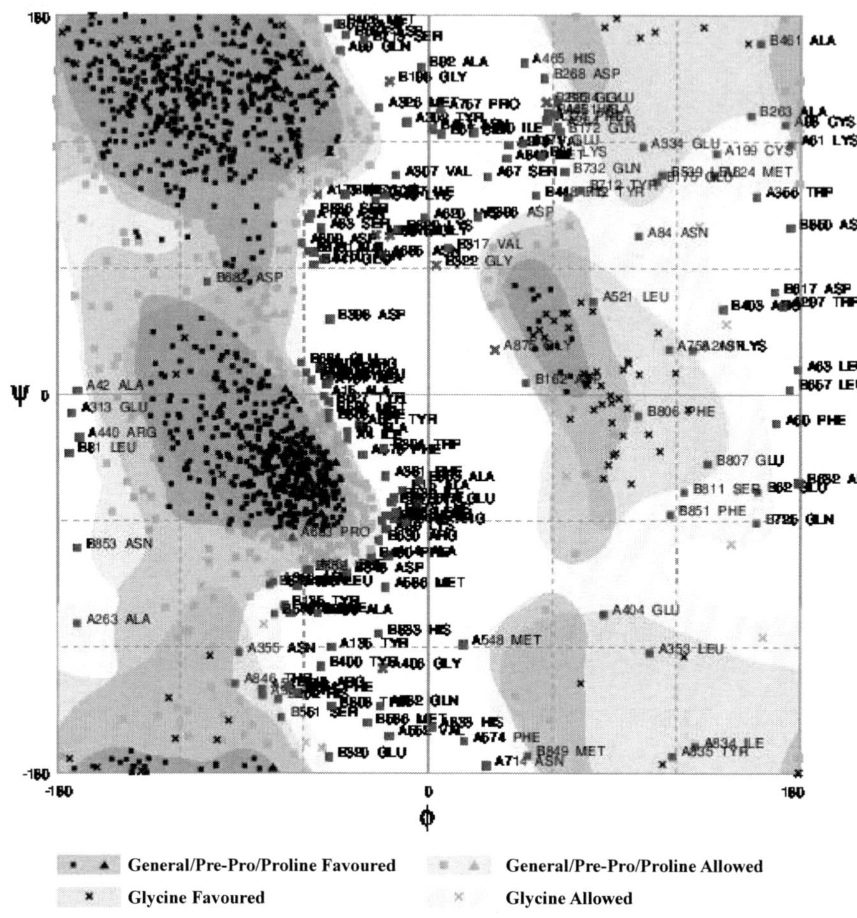

Figure 6.2a. Ramachandran plot of 3D model of homologous template μ-calpain (1QXP).

The verified 3D graph high score 0.43 for calpain10 (AAH07553) indicates that environment profile of the model is good (Figure 6.3). Profile score above zero in the Verify 3D graph [5, 17] corresponds to acceptable environment of the model. In Verified 3D plot, 25.81% of the residues had an averaged 3D-1D score >= 0.2.

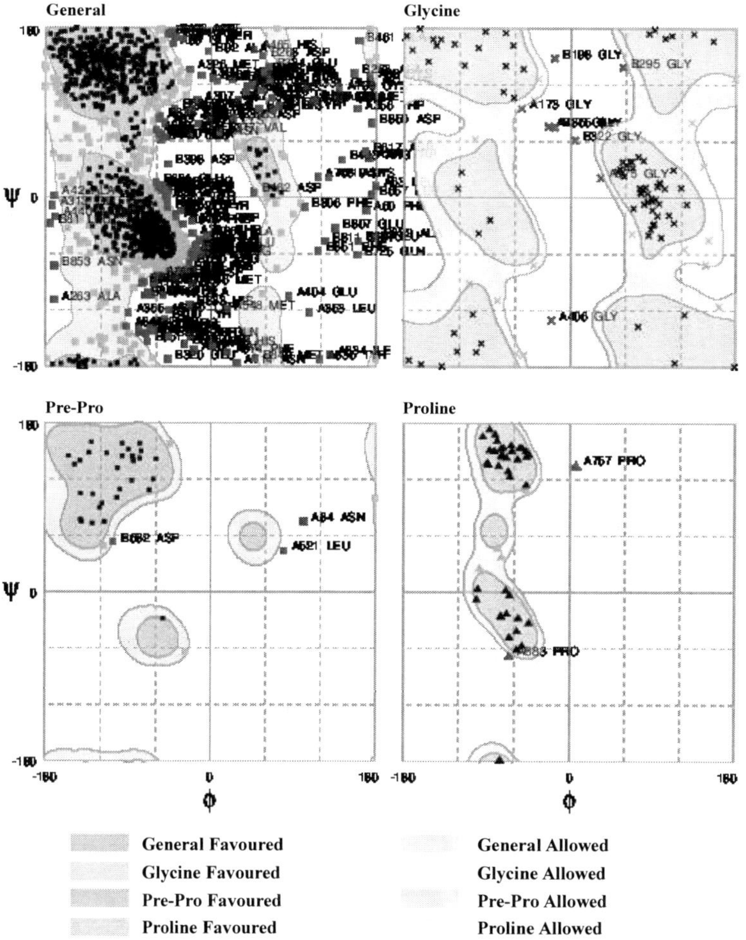

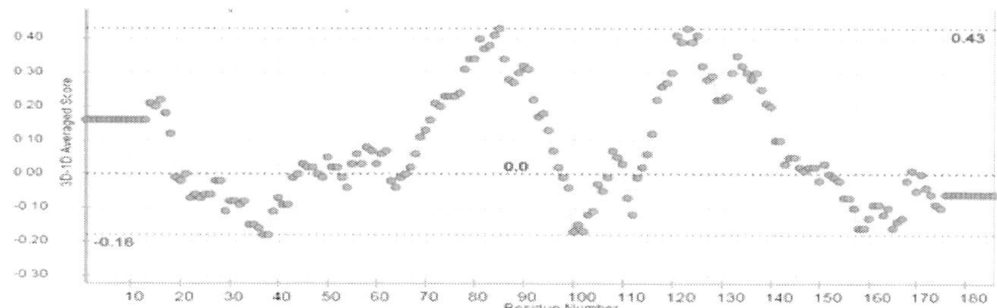

Figure 6.2b. Non-proline residues and non-glycine residue regions.

Figure 6.3. Verified 3D graph of calpain10 (AAH07553).

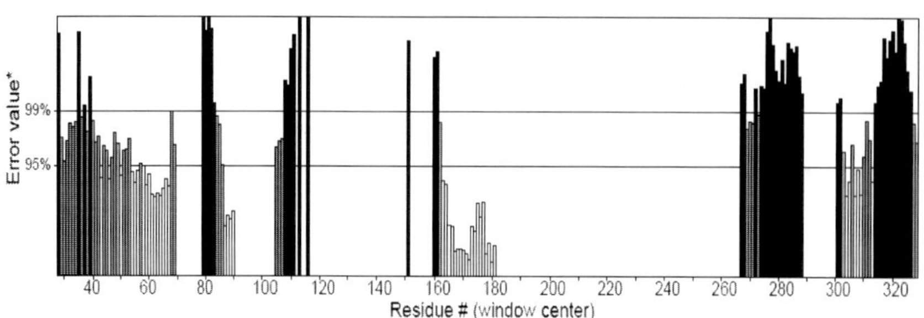

Figure 6.4. Errat graph of calpain10 (AAH07553).

The Errat is a program for verifying protein structures determined by crystallography. On the error axis, two lines are drawn to indicate the confidence with which it is possible to reject regions that exceed that error value [6]. Expressed as the percentage of the protein for which the calculated error value falls below the 95% rejection limit. Good high-resolution structures generally produce values around 95% or higher. For lower resolutions (2.5 to 3A) the average overall quality factor is around 9. The overall quality factor for calpain10 (AAH07553) was 33.333 (Figure 6.4).

6.2.3. Model Validation

Potential problems of protein structures based on energy plots are easily seen by ProSA and are displayed in a three-dimensional manner. ProSA was used to check the three-dimensional model of calpain10 (AAH07553) and μ-calpain (1QXP) proteins for potential errors (Figure 6.5 and Figure 6.6). The ProSA web z-scores of protein chains in PDB determined by X-ray crystallography (light blue) or NMR spectroscopy (dark blue) with respect to their length. The plot shows only chains with less than 1,000 residues and a z-score of 10. The z-scores of calpain10 (AAH07553) and μ-calpain (1QXP) are highlighted as large dots.

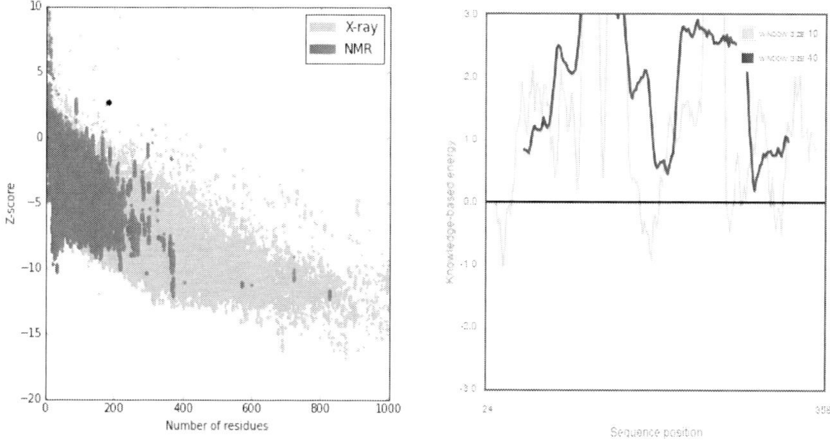

Figure 6.5. ProSA web service analysis of calpain10 (AAH07553) overall model qualit (a) and local model quality (b).

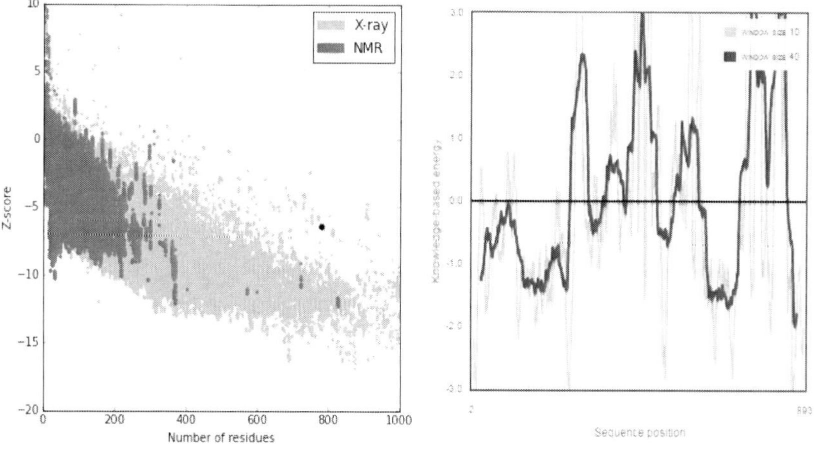

Figure 6.6. ProSA web service analysis of μ-calpain (1QXP) overall model quality (a) and local model quality (b).

The ProSA Z-score of -5.96 for calpain10 (AAH07553) [Figure 6.5] and -1.80 for μ-calpain (1QXP) protein indicates the overall model quality of target and template (Figure 6.6) measures the deviation of the total energy of the structure with respect to an energy distribution derived from random conformations [30].

                    Rajneesh Prajapat and Ijen Bhattacharya

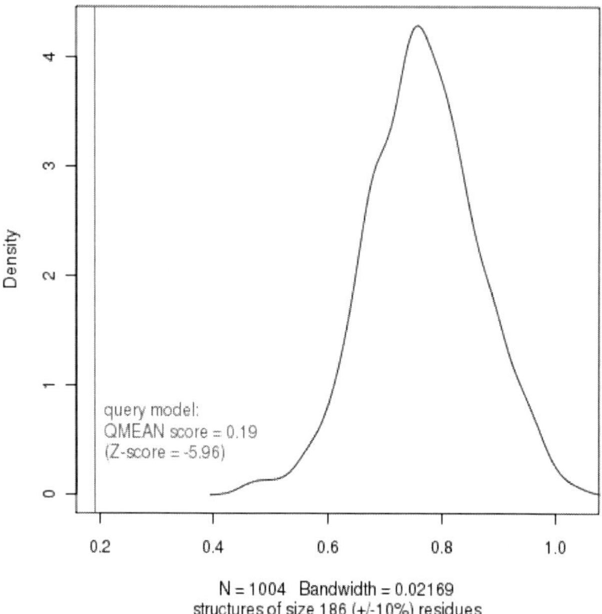

Figure 6.7a. The density plot for target calpain10 (AAH07553) showing the value of Z-score and QMEAN score.

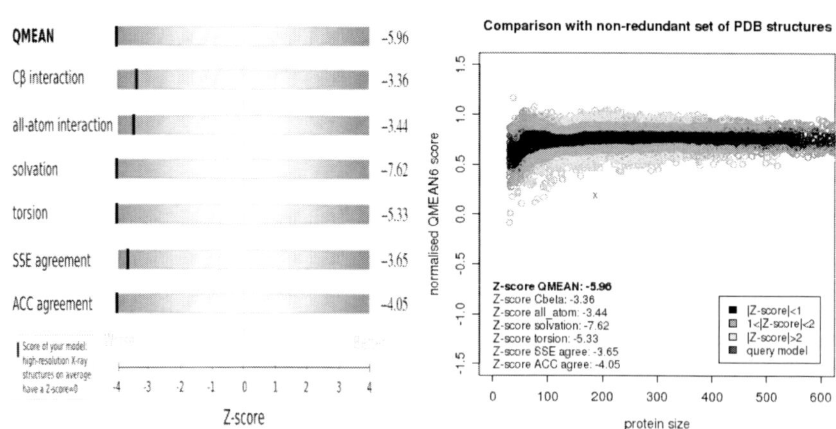

Figure 6.7b. Plot showing the QMEAN value as well as Z-score.

The QMEAN score of the calpain10 (AAH07553) model was 0.19 and the Z-score was -5.96 (Figure 6.7) and for the template μ-calpain (1QXP) was 0.60 with Z-score was -1.80 (Figure 8), which very closed to the value of 0 and this shows the fine quality of the both models [32; 24] because the estimated reliability of the model was expected to be in between 0 and 1 and this could be inferred from the density plot for QMEAN scores of the reference set. A comparison between normalized QMEAN score (0.40) and protein size in non-redundant set of PDB structures in the plot revealed different set of Z-values for different parameters such as C-beta interactions (-3.36), interactions between all atoms (-3.44), solvation (-7.62), torsion (-5.33), SSE agreement (-3.65) and ACC agreement (-4.05) (Figure 6.7; Table 2).

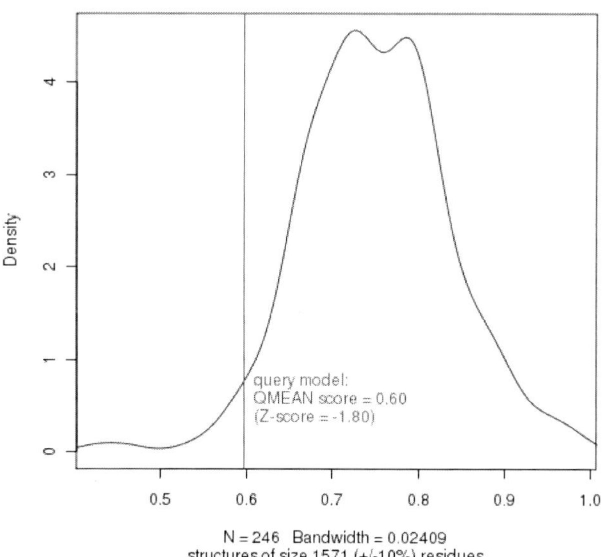

Figure 6.8a. The density plot for template μ-calpain (1QXP) showing the value of Z-score and QMEAN score.

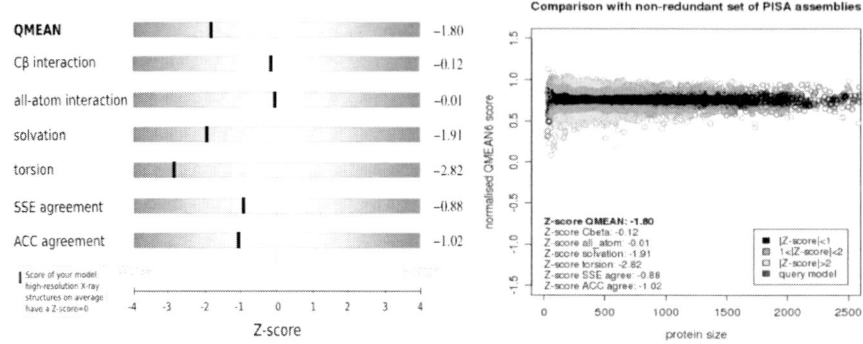

Figure 6.8b. Plot showing the QMEAN value as well as Z-score.

Figure 6.9. Calpain10 (AAH07553) ribbon model generated using QMEAN server. Estimated per-residue error visualized using a color gradient from blue (more reliable region) to red (potentially unreliable region).

Table 6.2. Table showing the QMEAN value as well as Z-score of calpain10 and μ-calpain

Accession No / PDB access code	Protein	Z - Scores						
		QMEAN	C-beta	All atom	Salvation	Torsion	SSE agree	ACC agree
AAH07553	calpain10	-5.96	-3.36	-3.44	-7.62	-5.33	-3.65	-4.05
1QXP	μ-calpain	-1.80	-0.12	-0.01	-1.91	-2.82	-0.88	-1.02

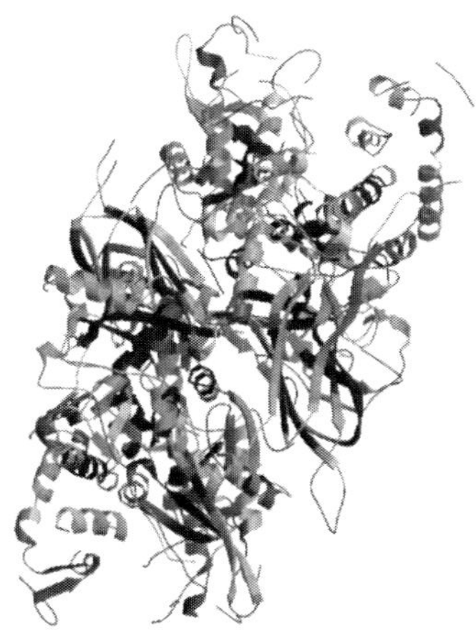

Figure 6.10. μ-calpain (1QXP) ribbon model generated using QMEAN server. Estimated per-residue error visualized using a color gradient from blue (more reliable region) to red (potentially unreliable region).

CONCLUSION

The human calpain10 (AAH07553) protein model was obtained through homology modelling and the main interactions are similar to those observed for template μ-calpain (1QXP). The calpain10 (AAH07553) model showed overall G-factor value of -0.65 with 55.9% residues in the favored region of Ramachandran plot and its template μ-calpain (1QXP) had 69.3% residues in favored region that indicates high accuracy of model predicted. The ProSA Z-score of -5.96 for calpain10 (AAH07553) and -1.80 for μ-calpain (1QXP) indicates the overall model quality of target and template measures the deviation of the total energy of the structure with respect to an energy distribution derived from random conformations. We hope these results will be useful for the design of inhibitors of calpain-10 and understanding of the mechanism of inhibition at the molecular level.

REFERENCES

[1]   Altschul SF, Madden TL, Schaffer AA, Zhang J, Zhang Z, Miller W, and Lipman DJ. 1997. "Gapped BLAST and PSI-BLAST: A new generation of protein database search programs." *Nucleic Acids Res.* 25: 3389-3402.

[2]   Arrington DD, Van Vleet TR, and Schnellmann RG. 2006. "Calpain 10: a mitochondrial calpain and its role in calcium-induced mitochondrial dysfunction." *Am. J. Physiol. Cell Physiol.* 291: 1159-1171.

[3]   Barry MM, Mol CD, Anderson WF, and Lee JS. 1994. "Sequencing and Modelling of anti-DNA immunoglobulin Fv domains. Comparison with crystal structures." *J. Biol. Chem.* 269: 3623–3632.

[4]   Benkert P, Künzli M, and Schwede T. 2009. "QMEAN server for protein model quality estimation." *Nucleic Acids Res.* 37: W510-4. doi: 10.1093/nar/gkp322., 2009.

[5]   Bowie JU, Luthy R, and Eisenberg D. 1991. "A method to identify protein sequences that fold into a known three-dimensional structure." *Science* 253: 164-170.

[6]   Colovos C, and Yeates TO. 1993. "Verification of protein structures: patterns of nonbonded atomic interactions." *Protein Sci.* 2 (9): 1511-9.

[7]   Cox NJ, Hayes MG, Roe CA, Tsuchiya T, and Bell GA. 2004. "Linkage of calpain 10 to type 2 diabetes: the biological rationale." *Diabetes* 53: S19-25.

[8]   Cuerrier D, Moldoveanu T, Inoue J, Davies PL, and Campbell RL. 2006. "Calpain inhibition by α-ketoamide and cyclic hemiacetal inhibitors revealed by X-ray crystallography." *Biochemistry* 45: 7446-7452.

[9]   Elbein SC, Chu W, Ren Q, Hemphill C, Schay J, Cox NJ, Hanis CL, and Hasstedt SJ. 2002. "Role of calpain-10 gene variants in

familial type 2 diabetes in Caucasians." *J Clin Endocrinol Metab.* 87: 650-654.

[10] Ghosh S, Watanabe RM, Hauser ER, Valle T, Magnuson VL, Erdos MR, Langefeld CD, Balow JJ, and Kohtamaki K, et al., 1999. "Type 2 diabetes: evidence for linkage on chromosome 20 in 716 Finnish affected sib pairs." *Proc Natl Acad Sci.* 96: 2198-2203.

[11] Goll DE, Thompson VF, Li H, Wei WEI, and Cong J. 2003. "The Calpain system." *Physiol. Rev.* 83 731-801.

[12] Heinrichs A. 2008. "Proteomics: Solving a 3D jigsaw puzzle." *Nat. Rev. Mol. Cell Biol.* 9: 3-3.

[13] Horikawa Y, Oda N, Cox NJ, Li X, Orho-Melander M, Hara M, Hinokio Y, Lindner TH, Mashima H, Schwarz PEH, Bosque-Plata L, Horikawa Y, Oda Y, Yoshiuchi I, Colilla S, Polonsky KS, Wei S, Concannon P, Iwasaki N, Schulze J, Baier LJ, Bogardus C, Groop L, Boerwinkle E, Hanis CL, and Bell GI. 2000. "Genetic variation in the gene encoding calpain-10 is associated with type 2 diabetes mellitus." *Nat. Genet.* 26: 163-175.

[14] Laskowski RA, Rullmann JA, MacArthur MW, Kaptein R, and Thornton JM. 1996. "AQUA and PROCHECK-NMR: Programs for checking the quality of protein structures solved by NMR." *J. Biomol. NMR.* 8: 477-486.

[15] Lima AH, Souza PRM, Alencar N, Lameira J, Govender T, Kruger HG, Maguire GEM, and Alves CN. 2012. "Molecular Modelling of *T. rangeli*, *T. brucei gambiense*, and *T. evansi* Sialidases in complex with the DANA inhibitor." *Chem. Biol. Drug Des.* 80: 114-120.

[16] Lovell SC, Davis IW, Arendall WB III, Bakker de PIW, Word JM, Prisant MG, Richardson JS, and Richardson DC. 2002. "Structure validation by Calpha geometry: phi, psi and Cbeta deviation." *Proteins: Structure, Function & Genetics* 50: 437-450.

[17] Luthy R, Bowie JU, and Eisenberg D. 1992. "Assessment of protein models with three-dimensional profiles." *Nature* 356: 83-85.

[18] Melo F, Devos D, Depiereux E, and Feytmans. 1997. "ANOLEA: a www server to assess protein structures." *Proc. Int. Conf. Intell. Syst. Mol. Biol.* 5: 187-190.

[19] Melo F, and Feytmans E. 1998. "Assessing protein structures with a non-local atomic interaction energy." *J. Mol. Biol.* 277: 1141-1152.

[20] Morris AL, MacArthur MW, Hutchinson EG, and Thornton JM. 1992. "Stereochemical quality of protein structure coordinates." *Proteins* 12: 345-364.

[21] Novotny WF, Girard TJ, Miletich JP, and Broze GJJ. 1998. "Platelets secrete a coagulation inhibitor functionally and antigenically similar to the lipoprotein associated coagulation inhibitor." *Blood* 72: 2020-2025.

[22] Ozaki T, Tomita H, Tamai M, and Ishiguro S. 2007. "Characteristics of mitochondrial calpains." *J. Biochem.* 142: 365-376.

[23] Picos-Cárdenas VJ, Sáinz-González E, Miliar-García A, Romero-Zazueta A, Quintero-Osuna R, Leal-Ugarte E, Peralta-Leal V, and Meza-Espinoza JP. 2015. "Calpain-10 gene polymorphisms and risk of type 2 diabetes mellitus in Mexican mestizos." *Genet Mol Res.* 14: 2205-2215.

[24] Prajapat R, Marwal A, and Gaur RK. 2014. "Recognition of Errors in the Refinement and validation of three-dimensional structures of AC1 proteins of begomovirus strains by using ProSA-Web." *J. of Viruses* ID 752656, .doi.org/10.1155/2014/752656, 2014.

[25] Resal R, and Ramteke PW. 2012. "Polymorphisms of Calpain 10 (Capn10) in Type 2 Diabetes mellitus - A Review." *Int. J. of Sci. Res.* 2: 1-4.

[26] Rawlings ND, Morton FR, and Barrett AJ. 2010. "MEROPS: the peptidase database." *Nucleic Acids Res.* 34: 270-272.

[27] Ridderstråle M, Parikh H, and Groop L. 2005. "Calpain 10 and type 2 diabetes: are we getting closer to an explanation"? *Curr. Opin. Clin. Nutr. Metab. Care.* 8: 361-6.

[28] Saez ME, Ramirez-Lorca R, Moron FJ, and Ruiz A. 2006. "The therapeutic potential of the calpain family: new aspects." *Drug Discov. Today* 1: 917-923.

[29] Sorimachi H, and Suzuki K. 2001. "The structure of calpain." *J. Biochemistry* 129: 653-664.

[30] Teilum K, Hoch JC, Goffin V, Kinet S, Martial JA, and Kragelund BB. 2005. "Solution structure of human prolactin." *J. of Mol. Bio.* 351:810–823.

[31] Weedon MN, Schwarz PE, Horikawa Y, Iwasaki N, Illig T, Holle R, Rathmann W, Selisko T, Schulze J, and Owen KR, et al., 2003. "Meta-analysis and a large association study confirm a role for calpain-10 variation in type 2 diabetes susceptibility." *Am. J. Hum. Genet.* 73(5): 1208-1212.

[32] Wiederstein M, and Sippl MJ. 2007. "ProSA-web: interactive web service for the recognition of errors in three-dimensional structures of proteins." *Nucleic Acids Research* 35: 407-410.

[33] Wiltshire S, Hattersley AT, Hitman GA, Walker M, Levy JC, Sampson M, O'Rahilly S, Frayling TM, Bell JI, and Lathrop GM, et al., 2001. "A genomewide scan for loci predisposing to type 2 diabetes in a U.K. population (the Diabetes UK Warren 2 Repository): analysis of 573 pedigrees provides independent replication of a susceptibility locus on chromosome 1q." *Am. J. Hum. Genet.* 69: 553-569.

[34] Zhang Z, Schwartz S, Wagner L, and Miller W. 2000. "A greedy algorithm for aligning DNA sequences." *J. Comput. Biol.* 7: 203-214.

In: Medical Bioinformatics and Biochemistry   ISBN: 978-1-53614-952-4
Editors: Rajneesh Prajapat et al.   © 2019 Nova Science Publishers, Inc.

Chapter 7

TO EVALUATE THE ROLE OF LEPTIN IN DIABETES FOR MALE AND FEMALE SUBJECTS

Jaipal Singh[1],, Ashish Sharma[1], Parduman Singh[2] and Rajneesh Prajapat[2]*
[1]Department of Medical Biochemistry,
Faculty of Medical Sciences, Geetanjali Medical College,
Geetanjali University, Udaipur, Rajasthan, India
[1]Department of Medical Biochemistry,
Saraswati Medical College, Unnao, Kanpur
[2]Department of Medical Biochemistry,
Faculty of Medical Sciences, Rama Medical College
and Hospital (NCR), Rama University, Kanpur, UP, India

---

* Corresponding Author's Email: jaypaul88@gmail.com.

ABSTRACT

The leptin promotes weight loss and can reverse diabetes by improving glucose tolerance by its action on hypothalamus. Present study will be conducted to evaluate the role of leptin in obesity associated maturity onset diabetes. This study was approved by ethical committee of the institution. Study was performed as a randomized controlled trial and with parallel design separately for male and female subjects. The serum glucose, HbA1c, cholesterol, triglycerides, HDL, LDL, insulin, TNF- α, adiponectin and leptin were analyzed by using semi auto analyzer (ERBA™), ELIZA assay kit and chemiluminescent assay. Study was done on 200 patients with type 2 diabetic patients to analyze the role of Leptin hormone in obesity induced type 2 diabetes. The type-II diabetes male patients were had significant higher fasting blood glucose ($P < 0.001$), significant high significant HbA1c ($P < 0.001$), non-significant high cholesterol ($P = 0.332$), non-significant high triglycerides ($P = 0.773$), significantly higher high-density lipoprotein ($P = 0.004$), significantly higher low-density lipoprotein ($P < 0.001$), significantly high insulin ($P < 0.001$), significantly high TNF- α ($P = <0.001$), significant low adiponectin ($P < 0.001$) and non-significant low leptin ($P < 0.001$) level were observed as compared to non-diabetic subjects.

The type-II diabetes female patients were had significant higher fasting blood glucose ($P < 0.001$), significant high significant HbA1c ($P < 0.001$), significant high cholesterol ($P = 0.016$), significant high triglycerides ($P = 0.025$), non-significantly higher high-density lipoprotein ($P = 0.599$), significantly higher low-density lipoprotein ($P < 0.001$), significantly high insulin ($P = <0.001$), significantly high TNF- α ($P = <0.001$), significant low adiponectin ($P < 0.001$) and significant low leptin ($P = 0.003$) level were observed as compared to non-diabetic subjects.

Keywords: leptin, HbA1c, ELIZA, TNF- α

INTRODUCTION

Diabetes is a metabolic disorder that causes hyperglycemia and giving rise to various vascular complications (Prajapat et al., 2017). Diabetes is a forthcoming epidemic all over the globe that caused due to ineffective secretion of insulin or insulin resistance (Awasthi et al.,

2016). Diabetes Mellitus is classified based on the pathogenic process that leads to hyperglycemia. The two mainly classified categories of DM include type 1 and type 2 DM (WHO, 1985).

Obesity and dyslipidemia take upper hand in the initiation, progression and complications of type 2 diabetes (Snehalatha et al., 2003; Kumar et al., 2017). Presently there are more than 62,000,000 people suffering from T2DM in India. Obesity and dyslipidemia are shown to play important role in its complications resulting in morbidity and mortality of T2DM (Kumar et al., 2017).

Obesity and type 2 diabetes are closely associated with low plasma levels of cytokine adiponectin in different ethnic groups of the society and indicate that the degree of hypoadiponectinemia is often more closely related to the degree of insulin resistance and hyperinsulinemia than to the degree of adiposity and glucose intolerance. (Weyer et al., 2001)

Leptin, the 167-amino acid protein, is a cytokine-like hormone secreted from white adipose tissue. It was the first adipocytokine identified, encoded by the ob gene. Leptin receptors are expressed in several different tissues. Adipocytes have been identified as the primary site for leptin expression, however it is also expressed in the gastric wall, vascular wall, placenta, ovary, in skeletal muscle, and the liver (Koerner et al., 2005; Brzozowski et al., 2005; Nawrot-Porabka et al., 2004; Konturek et al., 2004). Leptin has several roles, including growth control, metabolic control, immune regulation, insulin sensitivity regulation, and reproduction (Schwartz et al., 2000; Bajari et al., 2004; Kaur and Zhang 2005). However, its most important role is in body weight regulation.

The mechanisms involved in leptin secretion are all quite different. The rate of insulin-stimulated glucose utilization in adipocytes is a key factor linking leptin secretion to body fat mass (Friedman and Halaas, 1998). Although the mechanism is incompletely understood, it may involve glucose flux through the hexosamine pathway (Mueller et al., 1998). In addition, various observations indicate that leptin has a more important role than insulin in the CNS control of energy homeostasis. Insulin is secreted from the endocrine pancreas and exerts potent effects on peripheral nutrient storage. Insulin is an afferent signal to the CNS that causes long-term inhibitory effects on energy intake. Leptin receptors and insulin receptors are expressed by brain neurons involved in energy intake (Baskin and Wilcox, 1988; Baskin et al., 1999; Cheung et al., 1997), and administration of either peptide directly into the brain reduces food intake whereas deficiency of either hormone does the opposite (Sipols et al., 1995, Zhang et al., 1994).

Leptin is the chief regulator of the "brain gut axis," which provides a satiety signal through its action on the CNS receptors within the hypothalamus (Konturek et al., 2004). Activation of hypothalamic leptin receptors suppresses food intake and promotes energy expenditure pathways (Tilg and Moschen, 2004). Leptin levels decrease with weight reduction. The hypothesis that leptin resistance can occur in association with obesity was first suggested by the finding of elevated plasma leptin levels in obese humans (Mantzoros et al., 1998).

This hypothesis suggests that some cases of human obesity may be due to reduced leptin action in the brain, and affected individuals are unlikely to respond to pharmacological treatment with leptin. Several mechanisms contribute to leptin resistance. Leptin uptake into the brain is facilitated by leptin receptors expressed by endothelial cells (Guzik et al., 2003) in the bloodbrain barrier that function as leptin transporters. Impaired leptin transport across endothelial cells of the bloodbrain barrier is one potential mechanism leading to leptin resistance. Whether dysfunction of this transport process can lead to obesity remains to be determined, but it has been seen that in obese humans cerebrospinal fluids demonstrate low levels of leptin in comparison to plasma (Matarese et al., 2005). Upon activation of leptin receptors in the brain, a series of integrated neuronal responses required for food intake and energy balance are activated, and these neuronal effector pathways play a key role in energy homeostasis. Failure of one or more of these pathways in response to the leptin signalling will manifest as leptin resistance (Schwartz et al., 1997).

Reduced leptin-receptor signal transduction is another potential cause of leptin resistance. Like other cytokine receptors, activation of the leptin receptor induces expression of a protein that inhibits any further leptin signal transduction, termed 'suppressor of cytokine signaling-' (SOCS-3) (Guzik et al., 2005). The potential contribution of SOCS-3 to acquired forms of leptin resistance and obesity is an active area of study.

The present study was done on 200 male and female patients separately, with type 2 diabetic to analyze the role of leptin hormone. The serum glucose, HbA1c, cholesterol, triglycerides, HDL, LDL, insulin, TNF- α, adiponectin and leptin were analyzed.

7.1. METHODOLOGY

Study was performed in Goldfield Medical College, Faridabad, Haryana (India) and the project was approved by Geetanjali medical College, Geetanjali University, Udaipur [Rajasthan] INDIA, as a randomized controlled trial and with parallel design. According to ADA [2006], study was done on 200 male and female patients with type 2 diabetic to analyze the role of leptin hormone of age group range from 30 to 80 years were selected. Participants will be adults having obesity with Type 2 diabetes mellitus. Blood samples would be drawn to determine biochemical markers after taking consent from the patient.

Based on history, physical examination and preliminary lab investigations patients of obesity induced DM will be selected. Serum glucose (Myers et al., 2006) cholesterol (Kannel et al., 1979), triglycerides, HDL (Castelli et al., 1977), LDL (Nauck et al., 2002) and HbA1c (Jeppsson et al., 2002) will be performed on semi auto analyzer according to the methodology and instructions given on literature accompanying commercially available kits of ERBA company. Leptin along Adiponectin and tumor necrosis factor estimation will be done with help of ELIZA assay kit. Insulin will be done by chemiluminescent assay (Khoo et al., 2011).

All the markers mentioned above would be done from serum by collecting venous blood sample in the vacutainers. Blood sample would be withdrawn from antecubital vein. Subjects would be asked to have fasting of 8 to 12 hours. Results of biochemical markers would be analyzed to establish their role as predictor of obesity induced DM.

7.2. RESULTS

Based on the history, physical examination and preliminary lab investigations patients of obesity induced DM will be selected. In the past two decades there has been an explosive increase in the number of people diagnosed with DM particularly type 2 which is associated with modern lifestyle, abundant calories intake, reduced physical activity leading to obesity.

About 60-90% cases of type2 DM now appears to be related to obesity. Numerous studies have shown that insulin resistance precedes development of hyperglycemia in subjects that eventually develops type-2 DM. It has been realized that type2 DM develops only in insulin resistant subjects with the onset of beta cell dysfunction. Although Type 2 diabetes can be treated with oral hypoglycemic drugs for long time but ultimately, they require insulin to control their diabetes which has its side effects and available in inject able form only which is very cumbersome for the patient. Therefore, it is very necessary to look for other alternative therapy which has lesser side effects.

In this background the adipokine-leptin could be potential and beneficial alternative treatment modality. Leptin promotes weight loss, regulation of appetite and can reverse diabetes by improving glucose tolerance by its action on hypothalamus. More important leptin role in obesity energy homeostasis in relation to diabetes has received much attention. About 80% of type2 DM is overweight and in fact obesity is a primary risk factor for type 2 diabetes. Thus, a study is urgently required to explore the role of leptin in the etiology of obesity induced DM.

Table 7.1. Biochemical parameters of male subjects in diabetic patients and control group

Biochemical Markers	Male subjects (Mean ± SD)		t-value	P Value	Significance
	Diabetic patients	Non-Diabetic patients			
Fasting Blood Glucose [mg/dl]	183.8 ± 13.4	98.34 ± 1.94	60.89	<0.001	Significant
HbA1C [mmol/mol]	9.73 ± 0.53	6.4 ± 0.1	59.59	<0.001	Significant
Cholesterol [mg/dl]	183.56 ± 36.89	178.77 ± 30.79	0.97	0.332	Non-Significant
Triglycerides [mg/dl]	178.57 ± 80.08	175.47 ± 68.01	0.29	0.773	Non-Significant
High-density lipoprotein cholesterol [mg/dl]	39.55 ± 8.23	36.27 ± 7.25	2.92	0.004	Significant
Low-density lipoprotein cholesterol [mg/dl]	108.82 ± 33.40	91.50 ± 11.58	4.74	<0.001	Significant
Insulin [IU]	11.47 ± 8.32	6.45 ± 3.89	5.29	<0.001	Significant
TNF- α [ng/ml]	9.19 ± 8.61	2.52 ± 2.37	7.22	<0.001	Significant
Adiponectin [IU/ml]	9.96 ± 5.29	13.1 ± 4.2	-4.54	<0.001	Significant
Leptin [ng/ml]	35.26 ± 16.05	40.41 ± 23.98	-1.76	0.080	Non-significant

Values expressed as Mean ± SD, indicates p<0.001 (unpaired' test).

In the present study type-II diabetes male patients were had significant higher fasting blood glucose ($P < 0.001$), significant high significant HbA1c (P < 0.001), non-significant high cholesterol (P = 0.332), non-significant high triglycerides ($P = 0.773$), significantly higher high-density lipoprotein (P = 0.004), significantly higher low-density lipoprotein (P < 0.001), significantly high insulin (P < 0.001), significantly high TNF- α (P = <0.001), significant low adiponectin

($P < 0.001$) and non-significant low leptin ($P < 0.001$) level were observed as compared to non-diabetic subjects (Table 7.1).

Table 7.2. Biochemical parameters of female subjects in diabetic patients and control group

Biochemical Markers	Diabetic patients (Mean ± SD)		t-value	P Value	Significance
	Diabetic patients	Non-Diabetic patients			
Fasting Blood Glucose [mg/dl]	173.76 ± 9.28	91.32 ± 1.04	91.3	<0.001	Significant
HbA1C [mmol/mol]	7.9 ± 0.29	5.2 ± 0.02	96.08	<0.001	Significant
Cholesterol [mg/dl]	184.50 ± 40.73	172.99 ± 26.73	2.42	0.016	Significant
Triglycerides [mg/dl]	181.37 ± 76.63	158.28 ± 71.33	2.25	0.025	Significant
High-density lipoprotein cholesterol [mg/dl]	38.04 ± 8.45	37.45 ± 7.69	0.53	0.599	Non-significant
Low-density lipoprotein cholesterol [mg/dl]	10.47 ± 36.29	93.13 ± 11.51	4.69	<0.001	Significant
Insulin [IU]	12.14 ± 8.66	6.17 ± 2.95	6.73	<0.001	Significant
TNF- α [ng/ml]	8.39 ± 8.17	2.69 ± 2.65	6.85	<0.001	Significant
Adiponectin [IU/ml]	8.78 ± 4.09	12.01± 4.88	-5.16	<0.001	Significant
Leptin [ng/ml]	34.67 ± 16.77	42.60 ± 20.37	-3.06	0.003	Significant

Values expressed as Mean ± SD, indicates p<0.001 (unpaired 't' test).

In the present study type-II diabetes female patients were had significant higher fasting blood glucose ($P < 0.001$), significant high significant HbA1c (P < 0.001), significant high cholesterol (P = 0.016), significant high triglycerides ($P = 0.025$), non-significantly higher high-density lipoprotein (P = 0.599), significantly higher low-density lipoprotein (P < 0.001), significantly high insulin (P = <0.001), significantly high TNF- α (P = <0.001), significant low adiponectin ($P < 0.001$) and significant low leptin ($P = 0.003$) level were observed as compared to non-diabetic subjects (Table 7.2).

The leptin that secreted from adipose tissue does affect the insulin sensitivity and play a major role in pathogenesis of obesity related diabetes (Makki et al., 2013). The leptin plays an important role in energy homeostasis and administration help in regulating glucose homeostasis, improves glucose tolerance by enhancing insulin sensitivity (Denroche et al., 2012). According to new findings, leptin replacement therapy is a promising and safe strategy to treat type 1 and 2 diabetes (Kalra, 2013).

CONCLUSION

The diabetes is a metabolic disorder and forthcoming epidemic all over the globe that caused due to ineffective secretion of insulin. In the present study the type-II diabetes male patients were had significant higher fasting blood glucose ($P < 0.001$), significant high significant HbA1c ($P < 0.001$), non-significant high cholesterol ($P = 0.332$), non-significant high triglycerides ($P = 0.773$), significantly higher high-density lipoprotein ($P = 0.004$), significantly higher low-density lipoprotein ($P < 0.001$), significantly high insulin ($P < 0.001$), significantly high TNF-α ($P = <0.001$), significant low adiponectin ($P < 0.001$) and non-significant low leptin ($P < 0.001$) level were observed as compared to non-diabetic subjects. The type-II diabetes female patients were had significant higher fasting blood glucose ($P < 0.001$), significant high significant HbA1c ($P < 0.001$), significant high cholesterol ($P = 0.016$), significant high triglycerides ($P = 0.025$), non-significantly higher high-density lipoprotein ($P = 0.599$), significantly higher low-density lipoprotein ($P < 0.001$), significantly high insulin ($P = <0.001$), significantly high TNF-α ($P = <0.001$), significant low adiponectin ($P < 0.001$) and significant low leptin ($P = 0.003$) level were observed as compared to non-diabetic subjects.

Type 2 diabetes can be treated with oral hypoglycemic drugs but also require insulin to control diabetes which has its side effects on patient.

Therefore, it is necessary to look for other alternative therapy which has lesser side effects. The leptin plays an important role in energy homeostasis and administration help in regulating glucose homeostasis, improves glucose tolerance by enhancing insulin sensitivity. The leptin replacement therapy is a promising and safe strategy to treat type 1 and 2 diabetes.

REFERENCES

Ahima RS, and Osei SY. 2004. "Leptin signaling." *Physiology & Behavior* 81: 223-241.

Bajari TM, Nimpf J, and Schneider WJ. 2004. "Role of leptin in reproduction." *Curr Opin Lipidol.* 15: 315-319.

Baskin DG, Breininger JF, and Schwartz MW. 1999. "Leptin receptor mRNA identifies a subpopulation of neuropeptide Y neurons activated by fasting in rat hypothalamus." *Diabetes* 48: 828-833.

Baskin DG, Wilcox BJ, Figlewicz DP, and Dorsa DM. 1988. "Insulin and insulin-like growth factors in the CNS." *Trends Neurosci.* 11: 107-111.

Brzozowski T, Konturek PC, Konturek SJ, Brzozowska I, and Pawlik T. 2005. "Role of prostaglandins in gastroprotection and gastric adaptation." *J Physiol Pharmacol.* 56: 33-55.

Cheung CC, Clifton DK, and Steiner RA. 1997. "Proopiomelanocortin neurons are direct targets for leptin in the hypothalamus." *Endocrinology* 138: 4489-4492.

Denroche HC, Huynh FK, and Kieffer TJ. 2012. "The role of leptin in glucose homeostasis." *J of Diabetes Investigation* 3:115-129. doi:10.1111/j.2040-1124.2012.00203.x.

Friedman JM, and Halaas JL. 1998. "Leptin and the regulation of body weight in mammals." *Nature* 395: 763-770.

Guzik TJ, Bzowska M, and Kasprowicz A. 2005. "Persistent skin colonization with Staphylococcus aureus in atopic dermatitis: relationship to clinical and immunological parameters." *Clin Exp Allergy* 35: 448-455.

Guzik TJ, Korbut R, and Adamek-Guzik T. 2003. "Nitric oxide and superoxide in inflammation and immune regulation." *J Physiol Pharmacol.* 54: 469-487.

Kalra SP. 2008. "Central leptin insufficiency syndrome: an interactive etiology for obesity, metabolic and neural diseases and for designing new therapeutic interventions." *Peptides* 29 (1):127-138. doi:10.1016/j.peptides.2007.10.017.

Kalra SP. 2013. "Should leptin replace insulin as a lifetime monotherapy for diabetes type 1 and 2"? *Indian J of Endocr and Metabol.* 17:S23-S24. doi:10.4103/2230-8210.119496.

Kaur T, and Zhang ZF. 2005. "Obesity, breast cancer and the role of adipocytokines." *Asian Pac J Cancer Prev.* 6: 547-552.

Koerner A, Kratzsch J, and Kiess W. 2005. "Adipocytokines: leptin the classical, resistin, the controversical, adiponectin, the promising, and more to come." *Best Pract Res Clin Endocrinol Metab.* 19: 525-546 39.

Konturek PC, Brzozowski T, Burnat G, Kwiecien S, Pawlik T, Hahn EG, and Konturek SJ. 2004. "Role of brain-gut axis in healing of gastric ulcers." *J Physiol Pharmacol.* 55: 179-192.

Konturek SJ, Konturek JW, Pawlik T, and Brzozowski T. 2004. "Braingut axis and its role in the control of food intake." *J Physiol Pharmacol.* 55: 137-154.

Makki K, Froguel P, and Wolowczuk I. 2013. *"Adipose Tissue in Obesity-Related Inflammation and Insulin Resistance: Cells, Cytokines, and Chemokines.* ISRN Inflammation." 2013:139239. doi:10.1155/2013/139239.

Mantzoros CS, Frederich RC, Qu D, Lowell BB, MaratosFlier E, and Flier JS. 1998. "Severe leptin resistance in brown fat-deficient uncoupling protein promoter-driven diphtheria toxin A mice despite suppression of hypothalamic neuropeptide Y and circulating corticosterone concentrations." *Diabetes* 47: 230-238.

Matarese G, Moschos S, and Mantzoros CS. 2005. "Leptin in immunology." *J Immunol.* 174: 3137-3142

Mueller WM, Gregoire FM, Stanhope KL, Mobbs CV, Mizuno TM, Warden CH, Stern JS, and Havel PJ. 1998. "Evidence that glucose metabolism regulates leptin secretion from cultured rat adipocytes." *Endocrinology* 139: 551-558.

Nawrot-Porabka K, Jaworek J, Leja-Szpak A, Palonek M, Szklarczyk J, Konturek SJ, and Pawlik WW. 2004. "Leptin is able to stimulate pancreatic enzyme secretion via activation of duodeno-pancreatic reflex and CCK release." *J Physiol Pharmacol.* 55: 47-57.

Schwartz MW, Prigeon RL, Kahn SE, Nicolson M, Moore J, Morawiecki A, Boyko EJ, and Porte D. 1997. "Evidence that plasma leptin and insulin levels are associated with body adiposity via different mechanisms." *Diabetes Care* 20: 1476-1481.

Schwartz MW, Woods SC, Porte D Jr, Seeley RJ, and Baskin DG. 2000. "Central nervous system control of food intake." *Nature* 404: 661-671.

Sipols AJ, Baskin DG, and Schwartz MW. 1995. "Effect of intracerebroventricular insulin infusion on diabetic hyperphagia and hypothalamic neuropeptide gene expression." *Diabetes* 44: 147-151.

Tilg H, and Moschen AR. 2006. "Adipocytokines: mediators linking adipose tissue, inflammation and immunity." *Nat Rev Immunol* 6: 772-783.

Zhang Y, Proenca R, Maffei M, Barone M, Leopold L, and Friedman JM. 1994. "Positional cloning of the mouse obese gene and its human homologue." *Nature* 372: 425-432.

In: Medical Bioinformatics and Biochemistry   ISBN: 978-1-53614-952-4
Editors: Rajneesh Prajapat et al.   © 2019 Nova Science Publishers, Inc.

Chapter 8

STUDY OF LEPTIN AND ADIPONECTIN IN TYPE 2 DIABETES MELLITUS

Jaipal Singh[*] and Ashish Sharma

Department of Medical Biochemistry, Faculty of Medical Sciences,
Geetanjali Medical College, Geetanjali University,
Udaipur, Rajasthan, India

ABSTRACT

The diabetes mellitus is a metabolic disorder and forthcoming epidemic all over the globe that is caused due to insulin resistance. In this background the adipokine-leptin could be potential and beneficial alternative treatment modality. Leptin promotes weight loss, regulation of appetite & can reverse diabetes by improving glucose tolerance by its action on hypothalamus.

Present study will be conducted to evaluate the role of leptin in obesity associated maturity onset diabetes. Other biochemical markers such as triglycerides, total cholesterol, HDL, LDL, Adiponectin, tumor necrosis factor and insulin would also be studied as they are associated with obesity

---

[*] Corresponding Author's Email: jaypaul88@gmail.com.

of patient and insulin resistance. This study has been approved by ethical committee of the institution.

Study was performed as a randomized controlled trial and with parallel design. Physical examination and preliminary lab investigations patients of obesity induced DM will be selected. Serum glucose, cholesterol, triglycerides, HDL, LDL and HbA1c will be performed on semi auto analyzer according to the methodology & instructions given on literature accompanying commercially available kits of ERBATM.

Study was done on 200 patients with type 2 diabetic patients to analyze the role of Leptin hormone in obesity induced type 2 DM. In the present study type-II diabetes patients were had significantly higher fasting blood glucose ($P < 0.001$), glycosylated hemoglobin [HbA1c] ($P < 0.001$), cholesterol ($P = 0.015$) and high-density lipoprotein ($P = 0.009$), Low-density lipoprotein (<0.001) and non-significantly triglycerides (0.066) were observed. Further type-II diabetes patients were had significantly higher insulin ($P < 0.001$), TNF-α ($P < 0.001$), Leptin ($P < 0.001$) and high adiponectin ($P = 0.001$) were observed.

Although Type 2 diabetes can be treated with oral hypoglycemic drugs for long time but ultimately, they require insulin to control their diabetes which has its side effects & available in injectable form only which is very cumbersome for the patient. Therefore, it is necessary to look for other alternative therapy which has lesser side effects. Leptin together with other molecules that are secreted from adipose tissue does affect the insulin sensitivity & is accepted to play a major role in pathogenesis of obesity related diabetes.

Keywords: diabetes, adipokine, leptin

INTRODUCTION

Diabetes is a metabolic disorder that causes hyperglycemia and giving rise to various vascular complications [1]. Diabetes is a forthcoming epidemic all over the globe that caused due to ineffective secretion of insulin or insulin resistance [2]. Diabetes Mellitus is classified based on the pathogenic process that leads to hyperglycemia. The two mainly classified categories of DM include type 1 and type 2 DM [3].

Obesity and dyslipidemia take upper hand in the initiation, progression and complications of type 2 diabetes [4, 5]. Presently there are more than 62,000,000 people suffering from T2DM in India. Obesity and dyslipidemia are shown to play important role in its complications resulting in morbidity and mortality of T2DM [4]. Obesity and type 2 diabetes are closely associated with low plasma levels of cytokine adiponectin in different ethnic groups of the society and indicate that the degree of hypoadiponectinemia is often more closely related to the degree of insulin resistance and hyperinsulinemia than to the degree of adiposity and glucose intolerance [5]. Patients with Type 2 diabetes are associated with more than a two-fold excess mortality from cardiovascular disease, microvascular complications affecting the eyes, kidneys and nerves. If left untreated, these complications will lead to blindness, kidney failure, foot ulcers and finally leading to amputations of limbs [6].

Drugs have been targeting different defects of metabolism in diabetes patients, leaving the clinician with much better tools to tailor a more optimal treatment strategy towards diabetic patients. It is extremely important for medical professionals to have a proper knowledge and constantly updated scientific training in diabetes research, which will ultimately lead to the implementation of better and more cost-effective treatment and care programs for patients with diabetes [7].

Type 2 diabetes can be treated with oral hypoglycemic drugs for long time but ultimately, they require insulin to control their diabetes which has its side effects [8]. Therefore, it is necessary to look for other alternative therapy which has lesser side effects. Leptin together with other molecules that are secreted from adipose tissue does affect the insulin sensitivity and is accepted to play a major role in pathogenesis of obesity related diabetes [9].

Adipocyte derived proteins with anti-diabetic action include leptin, adiponectin, Omentin and Visfatin [10]. In this background the adipokine-leptin could be potential and beneficial alternative treatment modality. Leptin promotes weight loss, regulation of appetite and can

     Jaipal Singh and Ashish Sharma

reverse diabetes by improving glucose tolerance by its action on hypothalamus [11]. About 80% of type 2 DM patients are overweight and in fact obesity is a primary risk factor for type 2 diabetes. Thus, a study is urgently required to explore the role of Leptin in the etiology of obesity induced DM [12].

8.1. MATERIALS AND METHODS

Study was performed in Department of Biochemistry, in Goldfield Medical College, Faridabad, Haryana and the project was approved by Geetanjali medical College, Geetanjali University, Udaipur [Rajasthan] INDIA, as a randomized controlled trial and with parallel design. According to ADA [2006], the whole study comprised of 400 patients including 200 cases and 200 patients serving as control with type 2 diabetic of age group range from 30 to 80 years were selected.

Participants will be adults having obesity with Type 2 diabetes mellitus. Blood samples would be drawn to determine biochemical markers after taking consent from the patient. On the basis of history, physical examination and preliminary lab investigations patients of obesity induced DM will be selected. Serum glucose (13) cholesterol (14), triglycerides, HDL (15), LDL (16) and HbA1c (17) will be performed on semi auto analyzer according to the methodology and instructions given on literature accompanying commercially available kits of ERBA company.

Hormone Leptin along Adiponectin and tumor necrosis factor estimation will be done with help of ELIZA assay kit commercially available kits are based on the principle of ELIZA. Insulin will be done by chemiluminescent essay (18). All the markers mentioned above would be done from serum by collecting venous blood sample in the vacutainers. Blood sample would be withdrawn from antecubital vein. Subjects would be asked to have fasting of 8 to 12 hours. Results of biochemical markers would be analyzed to establish their role as predictor of obesity induced DM.

While analyzing the individual biochemical markers, we would consider the following value`s as their cut off upper limit.

- Cholesterol   - 210 mg/dl.
- Triglycerides   - 208 mg/dl.
- HDL   - 35 mg/dl
- LDL   - 80 mg/dl
- Leptin
- Adiponectin
- Tumor necrosis factor-α
- Insulin.

8.2. RESULTS

Blood samples of patients were drawn to determine biochemical markers after taking consent from the patients. Based on history, physical examination and preliminary lab investigations patients of obesity induced DM will be selected. Cholesterol, triglycerides, HDL, LDL was performed auto analyzer according to the methodology and instructions given on literature accompanying commercially available kits of ERBA^{TM}. Table 8.1 explains the ratio of male and female patients and Table 2 and 3 explain about past and present medical history of patients that were selected for study.

Table 8.1. Ratio of male and female patients selected for study

Gender	Diabetic patients (%)	Control group (%)	Total (%)
Male	99 (49.5%)	93 (46.5%)	192 (48%)
Female	101 (50.0%)	107 (53.5%)	208 (52%)
Total	200 (100%)	200 (100%)	400 (100%)

                    Jaipal Singh and Ashish Sharma

Table 8.2. Past/present history of disease of male and female of control group

Diseases	Control group		χ2-value	P Value	Significance
	MALE (%) N = 93	FEMALE (%) N = 107			
Anemia	25 (26.88%)	61 (57.01%)	17.22	<0.001	Significant
Hypertension	59 (63.44%)	78 (72.9%)	1.647	0.199	Non-Significant
Hypotension	52 (55.91%)	47 (43.93%)	2.401	0.121	Non-Significant

Table 8.3. Past/present history of disease of study subjects

Diseases	Diabetic patients (%) N = 200	Control group (%) N = 200	Total (%) N = 400	χ 2-value	P Value	Significance
Anemia	101 (50.5%)	86 (43.0%)	187 (46.75%)	1.968	0.161	Non-Significant
Hypertension	132 (66.0%)	137 (68.5%)	269 (67.25%)	5.926	0.015	Significant
Hypotension	100 (50.0%)	99 (49.5%)	199 (49.75%)	0.090	0.765	Non-Significant

The mean ± SD values of fasting blood glucose (FBG) was 178.78 ± 11.34 with observed significant P-value was <0.001. The mean ± SD values of HbA1C was 8.82 ± 0.41 with observed significant P-value was <0.001. The mean ± SD values of triglycerides was 179.98 ± 78.17 with non-significant P-value was <0.001, mean ± SD value of HDL was 39 ± 8.40 and LDL was 110 ± 34.8 with significant p-value (<0.001). The mean ± SD values of insulin was 11.81 ± 8.48 with significant P-value was <0.001. The mean ± SD values of insulin were 8. 79 ± 8.38 with significant P-value was <0.001.

The mean ± SD values of adiponectin 9.37 ± 5.13 with significant P-value was <0.001. The mean ± SD values of leptin 34.97 ± 16.38 with significant P-value was <0.001 (Table 8.4).

The observed t – value of adiponectin was -6.47 (Table 8.4), that indicate a low level of adiponectin is an independent risk factor for developing, metabolic syndrome and diabetes mellitus (19).

Table 8.4. Biochemical parameters of subjects with diabetes and the control group

Biochemical Markers	Diabetic patients Mean ± SD N = 200	Control group Mean ± SD N = 200	t-value	P-value	Significance
Fasting Blood Glucose [mg/dl]	178.78 ± 11.34	94.83 ± 1.49	103.80	<0.001	Significant
HbA1C [mmol/mol]	8.82 ± 0.41	5.8 ± 0.06	103.07	<0.001	Significant
Cholesterol [mg/dl]	184.04 ± 38.78	175.68 ± 70.16	2.44	0.015	Significant
Triglycerides [mg/dl]	179.98 ± 78.17	166.28 ± 70.16	1.84	0.066	Non-significant
High-density lipoprotein cholesterol [mg/dl]	39 ± 8.40	36.9 ±7.49	2.64	0.009	Significant
Low-density lipoprotein cholesterol [mg/dl]	110 ± 34.8	92.37 ± 11.35	6.81	<0.001	Significant
Insulin [IU]	11.81 ± 8.48	6.29 ± 3.42	8.54	<0.001	Significant
TNF- α [ng/ml]	8.79 ± 8.38	2.61 ± 2.52	9.99	<0.001	Significant
Adiponectin [IU/ml]	9.37 ± 5.13	12.52 ± 4.6	-6.47	<0.001	Significant
Leptin [ng/ml]	34.97 ± 16.38	41.51 ± 22.07	-3.36	<0.001	Significant

The mean ± SD values of fasting blood glucose (FBG) was 178.78 ± 11.34 with observed significant P-value was <0.001. The mean ± SD values of HbA1C was 8.82 ± 0.41 with observed significant P-value was <0.001. The mean ± SD values of triglycerides was 179.98 ± 78.17 with non-significant P-value was <0.001, mean ± SD value of HDL was 39 ± 8.40 and LDL was 110 ± 34.8 with significant p-value (<0.001). The mean ± SD values of insulin was 11.81 ± 8.48 with significant P-value was <0.001. The mean ± SD values of insulin were 8. 79 ± 8.38 with significant P-value was <0.001.

The mean ± SD values of adiponectin 9.37 ± 5.13 with significant P-value was <0.001. The mean ± SD values of leptin 34.97 ± 16.38 with significant P-value was <0.001 (Table 8.4).

The observed t – value of adiponectin was -6.47 (Table 9.4), that indicate a low level of adiponectin is an independent risk factor for developing, metabolic syndrome and diabetes mellitus (19).

Renju et al., (2012) reported that the fasting insulin, serum adiponectin levels and its correlation in patients with type 2 diabetics. Serum insulin and adiponectin levels were significantly decreased in patients compared to control subjects. In our study there is no significant correlation between adiponectin levels and insulin resistance in diabetic cases. (20).

Biochemical parameters of subjects with diabetes and control of the study are shown in Table 8.4. The type-II diabetes patients were had significantly higher fasting blood glucose ($P < 0.001$), glycosylated hemoglobin [HbA1c] ($P < 0.001$), cholesterol ($P = 0.015$) and high-density lipoprotein ($P = 0.009$), Low-density lipoprotein (<0.001) and non-significantly triglycerides (0.066) were observed. Further type-II diabetes patients were had significantly higher insulin ($P < 0.001$), TNF-α ($P < 0.001$), adiponectin ($P = 0.001$) and Leptin ($P < 0.001$) were observed [Table 8.4].

CONCLUSION

The diabetes is a metabolic disorder and forthcoming epidemic all over the globe that caused due to ineffective secretion of insulin. In the present study type-II diabetes patients were had significantly higher fasting blood glucose ($P < 0.001$), glycosylated hemoglobin [HbA1c] ($P < 0.001$), cholesterol ($P = 0.015$) and high-density lipoprotein ($P = 0.009$), Low-density lipoprotein (<0.001) and non-significantly triglycerides (0.066) were observed. Further type-II diabetes patients were had significantly higher insulin ($P < 0.001$), TNF- α ($P < 0.001$), Leptin ($P < 0.001$) and high adiponectin ($P = 0.001$) were observed. The observed t – value of adiponectin was -6.47 (Table 4), that indicate a low level of adiponectin is an independent risk factor for developing, metabolic syndrome and diabetes mellitus. Type 2 diabetes can be treated with oral hypoglycemic drugs but also require insulin to control diabetes which has its side effects on patient. Therefore, it is necessary to look for other alternative therapy which has lesser side effects. Leptin together with other molecules that are secreted from adipose tissue does affect the insulin sensitivity and is accepted to play a major role in pathogenesis of obesity related diabetes.

REFERENCES

[1]   Prajapat R, Bhattacharya I, and Jakhalia A. 2017. "Combined Effect of Vitamin C and E Dose on Type 2 Diabetes Patients." *Adv in Diabetes and Metabolism* 5: 21 - 25. doi: 10.13189/adm.2017.050201.

[2]   Awasthi A, Parween N, Singh VK, Anwar A, Prasad B, and Kumar J. 2016. "Diabetes: Symptoms, Cause and Potential Natural Therapeutic Methods." *Adv. in Diabetes and Metabolism* 4: 10-23. doi: 10.13189/adm.2016.040102.

[3]   World Health Organization: *Diabetes mellitus, Technical Report Series 727*, WHO, Geneva. 1985.

[4]   Kumar PA, Pandey D, and Pandit A. 2017. "Obesity and Lipid Profile Study in Type 2 Diabetes Patients with Auditory and Reaction Time Deficits and Non-Diabetic Control Subjects." *Adv in Diabetes and Metabolism* 5: 1-5.

[5]   Snehalatha C, Viswanathan V, and Ramachandran A. 2003. "Cutoff values for normal anthropometric variables in Asian Indian adults." *Diabetes Care* 26:1380-4.

[6]   Snehalatha C, Viswanathan V, and Ramachandran A. 2003. "Cutoff values for normal anthropometric variables in Asian Indian adults." *Diabetes Care* 26:1380-4.

[7]   Green LW, Brancati FL, and Albright A. 2012. "The Primary Prevention of Diabetes Working Group. Primary prevention of type 2 diabetes: integrative public health and primary care opportunities, challenges and strategies." *Family Practice* 29:i13-i23. doi:10.1093/fampra/cmr126.

[8]   Noureddine H, Nakhoul, N, Galal A, Soubra L, and Saleh M. 2014. "Level of A1C control and its predictors among Lebanese type 2 diabetic patients." *Therapeutic Adv in Endo and Metabol.* 5: 43–52.

[9]   Mohamed-Ali V, Pinkney JH, and Coppack SW. 1998. "Adipose tissue as an endocrine and paracrine organ." *Int J Obes Relat Metab Disord.* 22:1145-1158.

[10]   Masur K, Thevenod F, and Zanker KS. 2008. "Diabetes and cancer. Epidemiological Evidence and Molecular links." *Front Diabetes* 19.

[11]   Paz-Filho G, Mastronardi C, Wong ML, and Licinio J. 2012. "Leptin therapy, insulin sensitivity, and glucose homeostasis." *Indian J of Endo and Metabol.* 16:S549-S555. doi:10.4103/2230-8210.105571.

[12]   Al-Goblan AS, Al-Alfi MA, and Khan MZ. 2014. "Mechanism linking diabetes mellitus and obesity." *Diabetes, Metabolic Syndrome and Obesity: Targets and Therapy* 7: 587-591. doi:10.2147/DMSO.S67400.

[13]   Myers GL, Miller WG, Coresh J, Fleming J, Greenberg N, Greene T, Hostetter T, Andrew SL, Panteghini M, Welch M, and Eckfeldt JH. 2006. "Recommendations for Improving Serum Creatinine Measurement: A report from laboratory working group of the National kidney disease education program." *Clinical Chemistry* 52: 5 – 18.

[14] Kannel WB, Castelli WP, and Gordon T. 1979. "Cholesterol in the prediction of atherosclerotic disease; new perspectives based on the Framingham study." *Ann Intern Med.* 90:85.

[15] Castelli WP, Doyle JT, Gordon T, Hames CG, Hjortland MC, Hulley SB, Kagan A, and Zukel WJ. 1977. "HDL cholesterol and other lipids in coronary heart disease. The Cooperative Lipoprotein Phenotyping Study." *Circulation* 55: 767-772.

[16] Nauck M, Warnick GR, and Rifai N. 2002. "Methods for Measurement of LDL-Cholesterol: A Critical Assessment of Direct Measurement by Homogeneous Assays versus Calculation." *Clinical Chemistry* 48: 236-254.

[17] Jeppsson JO, Kobold U, Barr J, Finke A, Hoelzel W, Hoshino T, Miedema K, Mosca A, Mauri P, Paroni R, Thienpont L, Umemoto M, and Weykamp C. 2002. "Approved IFCC reference method for the measurement of HbA1c in human blood." *Clin Chem Lab Med.* 40:78-89.

[18] Khoo CM, Sairazi S, and Taslim S, et al., 2011. "Ethnicity Modifies the Relationships of Insulin Resistance, Inflammation, and Adiponectin With Obesity in a Multiethnic Asian Population." *Diabetes Care* 34:1120-1126. doi:10.2337/dc10-2097.

[19] Nedvídková J, Smitka K, Kopský V, and Hainer V. 2005. "Adiponectin, an adipocyte-derived protein." *Physiol Res.* 54 (2): 133–40.

[20] Fruhbeck, G, Gomez-Ambrosi, J, Muruzabal FJ, and Burrell MA. 2001. *Am. J. Physiol. Endocrinol. Metab.* 280: E827–E847.

In: Medical Bioinformatics and Biochemistry   ISBN: 978-1-53614-952-4
Editors: Rajneesh Prajapat et al.   © 2019 Nova Science Publishers, Inc.

Chapter 9

STUDY ON CORRELATION OF INFLAMMATORY MARKERS ON TYPE-II DIABETES

Krattika Singhal[*] *and Sandesh Shrestha*
Department of Medical Biochemistry,
Faculty of Medical Sciences,
Rama Medical College, Hapur, UP, India

ABSTRACT

Type-II Diabetes, a metabolic disease, presenting with hyperglycemia resulting from insulin deficiency or decreased glucose utilization or increased glucose production has been found to have an association with various inflammatory markers. To determine and compare the serum level of C-reactive protein and uric acid and also to observe the association between them.

The mean values of C-reactive protein and uric acid obtained from 60 subjects, were analyzed by student't' test and Pearson's correlation. The

---

[*] Corresponding Author's Email: sandycrestha@yahoo.com.

                     Krattika Singhal and Sandesh Shrestha

mean level of C-reactive protein activity of diabetic patients is significantly (p=0.028) higher as compared to normal subjects and the mean serum uric acid is also apparently higher, however, a significant difference could not be established (p=0.072). C-reactive protein and uric acid showed insignificant association to the Diabetic mellitus. Also, had an insignificant correlation with each other.

Keywords: diabetes, C-reactive protein, uric acid

INTRODUCTION

India accounts for the largest number of people suffering from Diabetes Mellitus in the world followed by China and the United States. The incidence of Diabetes Mellitus worldwide was estimated to be 2.8% in year 2000 and is expected to increase up to 4.4% in 2030. The number of Diabetes Mellitus patients is expected to be 79.4 million by 2030 in India [1, 2].

Diabetes Mellitus is a group of metabolic diseases presenting with hyperglycemia resulting from insulin deficiency or decrease glucose utilization or increase glucose production. Previous studies have suggested that serum CRP and uric acid levels are positively associated with the development of type II Diabetes [3, 4].

9.1. URIC ACID AS AN INFLAMMATORY MARKER IN DIABETES MELLITUS

The association between the blood glucose and the serum Uric acid levels has been known for quite some time (There have been considerable studies showing an association between blood glucose and serum uric acid. In addition, an affirmative correlation has been established between the serum Uric acid levels and the progression of type II Diabetes Mellitus [5, 6]. In an individual with impaired glucose tolerance, an elevated serum uric acid level was found to increase the risk of developing type II Diabetes Mellitus [7].

Although hyperuricemia is considered as a risk factor for type II Diabetes Mellitus, the causal association between them is contradictory. In spite the fact that many studies have demonstrated an association between high serum uric acid and insulin resistance, the causal effect between high serum uric acid and insulin resistance, is not yet clarified. Uric acid – the product of purine catabolism – was found to be associated with hypertension, obesity, dyslipidemia as well as hyperinsulinemia [8].

Apart from the established casual association of hyperuricemia leading to gout and metabolic syndrome leading to Diabetes Mellitus, both hyperuricemia and metabolic syndrome are related to hyperinsulinemia. The patients with both gout and type II Diabetes Mellitus diseases exhibit a mutual inter-dependent effect on higher incidences. The association is quite intricate and insulin resistance seems to be possibly just a common link [9].

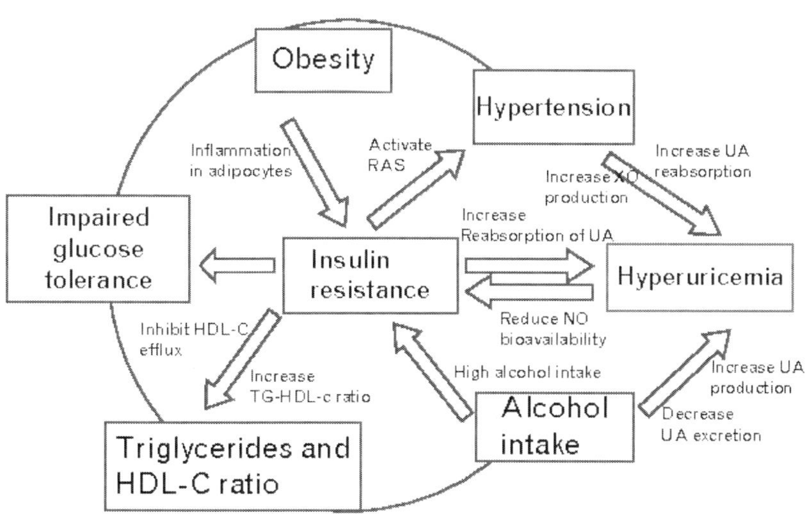

Figure 9.1. The image demonstrates the interaction between the components of metabolic syndrome, insulin resistance, and hyperuricemia.

Enhanced mean serum uric acid levels were found to be significantly increased due metabolic syndrome, and furthermore, the prevalence of metabolic syndrome also increased significantly with uric acid levels [10-12]. The highest quartile of uric acid levels, the risks were substantially higher for metabolic syndrome compared with those in the lowest quartile of uric acid levels. This demonstrates that higher uric acid levels are associated with metabolic syndrome, and the converse is also true, that patients with hyperuricemia frequently have metabolic syndrome. Moreover, Facchini et al. [8] had suggested that insulin resistance is the pathophysiological mechanism for the association.

The uric acid blocked acetylcholine-mediated arterial dilation, suggesting that uric acid can impair endothelial function [13]. The endothelial nitric oxide synthase deficiency results in the features of insulin resistance and metabolic syndrome. Because uric acid has been shown to reduce nitric oxide bioavailability [14, 15] and reducing endothelial nitric oxide supply is a known mechanism for inducing insulin resistance [16]. Markkola and Jarvinen indicated that serum uric acid level is inversely correlated with insulin sensitivity, and uric acid was suggested to play an important role in the function of the β – cell in the patients with type II diabetes mellitus even in states prior to hyperuricemia [17, 18]. A large epidemiologic study also showed that high serum uric acid levels had a positive correlation with fasting serum insulin levels, and Tsouli et al., have reviewed the association between elevated uric acid and insulin resistance [19, 20].

A person with Diabetes Mellitus has a two-fold greater risk of suffering mayo-cardial infraction that does a non-diabetic person of the same age and sex. Insulin resistance or glucose intolerance is a component of a cluster of metabolic risk found in individuals who are prone to cardiovascular diseases. This cluster also includes obesity (especially weight gain in the abdominal region), atherogenic dyslipidemia, hypertension, elevated fibrinogen or plasminogen activator inhibitor-1, and inflammatory state that indicated by elevated C-reactive protein and uric acid. This widely prevalent condition is called metabolic syndrome [21-23].

9.2. CRP AS AN INFLAMMATORY MARKER IN DIABETES MELLITUS

C- reactive protein is an annular (ring-shaped), a pentameric protein found in the blood plasma, the levels of which rise in response to inflammation (i.e., C-reactive protein is an acute-phase protein of hepatic origin that increases following Interleukin-6 secretion from macrophages and T-cell). Its physiological role is to bind to lysophosphatidylcholine expressed on the surface of dead or dying cells in order to activate the complement system via the C1Q complex [24]. CRP is synthesized by the liver in response to factors released by macrophages and fat cells (adipocytes) [25].

CRP binds to phosphocholine expressed on the surface of dead or dying cells and some bacteria. This activates the complement system, promoting phagocytosis by macrophages, which clears necrotic and apoptotic cells and bacteria. This so-called acute phase response occurs as a result of rising in the concentration of IL-6, which is produced by macrophages [26] in response to a wide range of acute and chronic inflammatory conditions such as bacterial, viral or fungal infections; rheumatic and other inflammatory diseases; malignancy; and tissue injury and necrosis. These conditions cause the release of IL-6 and other cytokines that trigger the synthesis of CRP and fibrinogen by the liver.

Hyperglycemia is known to stimulate the release of the inflammatory cytokines TNF- α and IL-6 from various cell types. Hyperglycemia can result in the induction and secretion of acute phase reactants by adipocytes. Prolonged exposure to hyperglycemia is now recognized as the primary causal factor in the pathogenesis of diabetic complication s including atherosclerosis in monocytes, chronic hyperglycemia causes a dramatic increase in the release of cytokines. Hyperglycemia directly examined the association between CRP and fasting insulin, fasting glucose and insulin resistance [27].

Diabetic individuals have higher concentrations of CRP and it closely related to adiposity. The increased level of serum CRP in obese individuals is due to increased secretion of IL-6 and TNF –α in adipocytes, which regulate CRP production in hepatocytes and induce a chronic inflammatory state.

Elevated levels of CRP and plasminogen-activator inhibitor (PAI) have been demonstrated to predict the incidence of type II diabetes mellitus. Abdominal obesity and the subsequent secretion of proinflammatory cytokines and acute phase reactants may contribute to the relationship between chronic inflammation and type II diabetes mellitus. Adipocytes (fat cells) that secrete IL-6 and the amount of IL-6 produced by adipocytes are proportional to the amount of fat cell mass [28].

Pradhan et al. [29] suggested that patients with elevated basal levels of CRP are at an increased risk of type II diabetes mellitus. It is well-known that cytokines operate as a network in stimulating the production of acute-phase proteins like CRP. In vivo studies have shown that adipose tissue secretes IL-6, which regulates CRP production and could, potentially, induce chronic systemic inflammation in subjects with excess body fat [30].

9.3. METHODS

9.3.1. Sample Size and Sampling method

This study was conducted in Teerthanker Mahaveer Medical College and Research Center, Moradabad, U.P., India. In the present study, 60 subjects participated, out of which 30 were patients aged 35 to 75 years who were diagnosed as diabetics and confirmed by the estimation of fasting serum glucose (> 126 mg/dl) on two occasions were selected from the Medicine OPD and IPD. 30 Normal healthy subjects, age, and sex matched with the diabetic patients were selected as controls. After overnight fasting for 8 hours, about 5ml of venous blood was drawn with aseptic precaution from the antecubital vein of all the subject and dispensed into following vials for various biochemical tests:

- Fluoride oxalate vial for fasting plasma glucose (FPG) estimation.
- Plain vial for C –reactive protein and Uric acid estimation.
- *Fasting plasma glucose was estimated by the* GOD-POD *method:* Glucose is oxidized by glucose oxidase (GOD) to produce gluconic acid and hydrogen peroxide. The hydrogen peroxide then reacts with 4-aminoantipyrine (4AAP) and 4-hydroxybenzoic acid (4HBA) in presence of peroxidase (POD) enzyme to form a pink colored quinonimine dye. The intensity of the pink color is directly proportional to the concentration of glucose present and O.D. is measured at 505nm [31].
- *Estimation of serum Uric acid by uricase method:* Uric acid is oxidized to allantoin by uricase with the production of hydrogen peroxide. The peroxide then reacts with 4-aminoantipyrine and TOOS in presence of peroxidase to yield quinoneimine dye. The absorbance of this dye is at 546 nm which is proportional to uric acid in the sample [32].
- *Estimation of serum CRP with turbidimetric immunoassay:* The test specimen is mixed with activation buffer (R1), TURBILYTE-CRPTM latex reagent (R2) and allowed to react. The presence of CRP in the test specimen results in the formation of an insoluble complex producing a turbidity, which is measured at 546 nm wavelength. The increase in turbidity corresponds to the concentration of CRP in the test specimen [33].

                    Krattika Singhal and Sandesh Shrestha

- *Inclusion and exclusion criteria:* Individuals diagnosed with Diabetes Mellitus by estimation of Fasting Plasma Glucose (FPG) $\geq$126mg/dl on two occasions were included. Individuals suffering from other inflammatory diseases like Tuberculosis, Leprosy and pregnancy, Cancer, Skin diseases, Gout, Liver, and Kidney diseases were excluded to rule out any increase in inflammatory markers due to other causes.

- *Statistical analysis and software used:* The data obtained were analyzed via IBM SPSS version 21 to determine the student 't' and Pearson's correlation coefficient.

- *Result:* In the present study, a total of 60 subjects participated out of which 30 were Diabetes patient and 30 were age and sex-matched healthy nondiabetic individuals. The study revealed, out of 30 patients, 17 patients i.e., about 56% had an increased level of CRP from the reference range (1-5mg/dl). While 14 patients i.e., about 46% had increased level of Serum uric acid from the reference range (4-8 mg/dl) and 7 patients i.e., about 23% had increased both serum CRP as well as serum uric acid level.

Diabetes is a metabolic condition with the hyperglycemic state with reduced insulin activity or lowered glucose intolerance giving rising to several clinical conditions like metabolic syndrome, atherosclerosis, cataract, kidney failure and cardiovascular complications. Previous

Table 9.1. Comparison of fasting plasma glucose, serum uric acid and serum CRP level between study groups

Parameters	Diabetic Mean ± S.D.	Non-Diabetic Mean ± S.D.	p-value
FBG (mg/dl)	199 ± 90.78**	97.71 ± 13	0.00
Uric acid (mg/dl)	6.45 ± 0.86	4.8 ± 0.83	0.072
CRP (mg/dl)	5.88 ± 2.66*	3.90 ± 1.2	0.028

* Highly significant at p < 0.01.

** Significant at p < 0.05.

Table 9.2. Comparison of fasting plasma glucose, serum uric acid and CRP level between study groups

VARIABLE (n=30)	URIC ACID	C- Reactive Protein
Fasting Blood Sugar r	-0.082	-0.025
p	0.669*	0.896*
Uric acid r	-	-0.291
p		0.119*

* Highly significant at p < 0.01.

** Significant at p < 0.05.

⁺In- significant at p > 0.05.

r: Pearson's correlation; n: number of patient.

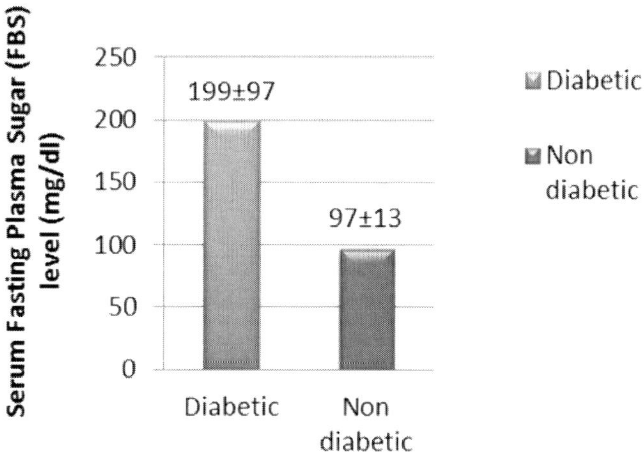

Figure 9.2. Comparison of Fasting Plasma Sugar between subjects.

studies have suggested that serum CRP and Uric acid levels are positively associated with the development of type II Diabetes Mellitus [3, 4]. However, our study had a contrast result, with the insignificant negative association of hyperglycemia with C- reactive protein and uric acid. Generally, with hyperglycemia rise in oxidative stress has been observed which have been associated with inflammation with enhanced pro-inflammatory cytokines [32, 34]. The slight negative association of this hyperglycemia with uric acid could be explained by the dual nature of uric acid i.e., acting either antioxidant or oxidant [16]. However, the insignificant slight negative association of uric acid with C-reactive protein questions whether uric acid should be seen as an inflammatory marker.

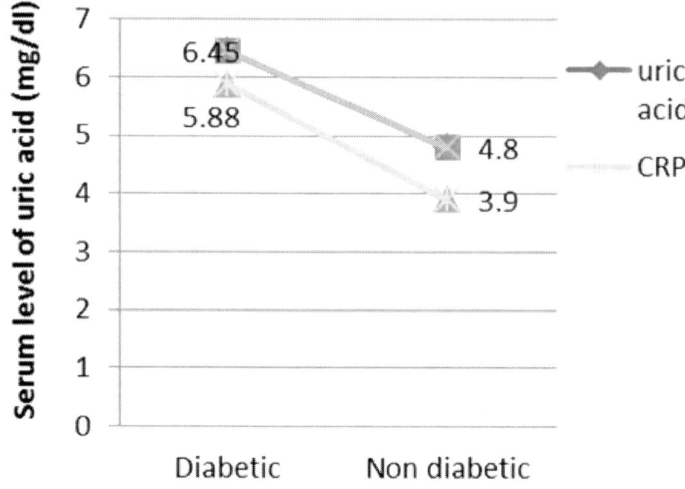

Figure 9.3. A comparison study between Patients and Control Group.

CONCLUSION

An attempt to see if uric acid acts as an inflammatory parameter was not sufficed due to its insignificant association to C-reactive protein, which is a potent marker of the inflammatory state. Our study has provided an evident that C-reactive protein might not have a significant role and rise in the diabetic state although this finding could go contradictory due to small sample size and short duration of the study.

REFERENCES

[1]   Wild S, Roglic G, Green A, Sicree R, and King H. 2004. "Global Prevalence of Diabetes." *Diabetes Care* 27(5):1047-1051.

[2] Mohan V, Sandeep S, Deepa R, Shah B, and Varghese C. 2007. "Epidemiology of type II diabetes: Indian Scenario." *Ind J Med Res*. 125:217-230.

[3] Dehghan A, Kardys I, Maat MPD, Uitterlinden AG, Sijbrands EJG, Bootsma AH, Stijnen T, Hofman A, Schram MT, and Wittermen JCM. 2007. "Genetic variation, C-reactive protein levels, and incidence of diabetes." *Diabetes* 56: 872-878.

[4] Kodam S, Satio K, Yachi Y, Asumi M, Sugawara A, Totsuka K, Saito A, and Sone H. 2009. "Association between Serum Uric acid and development of type II diabetes." *Diabetes Care* 32: 1737-1742.

[5] Herman JB, Medalie JH, Goldbourt U. 1976. *"Diabetes, pre diabetes and uricemia. Diabetologia."* 12 :47-52.

[6] Kodama S, Satio K, Yachi Y, Asumi M, Sugawara A, and Totsuka K, et al., 2009. "The between serum uric acid and the development of type II Diabetes Mellitus - A meta-analysis. *Diabetes Care* 32:1737-42.

[7] Caroline K, Denise V, Simerjot K, and Elizabeth B. 2009. "The serum uric acid levels improve the prediction of the incident of the incident type II Diabetes in individuals with impaired fasting glucose levels. The Rancho Bernardo Study." *Diabetes Care* 2:1272-73.

[8] Meshkani R, Zargari M, and Larijiani B. 2011. "The relationship between uric acid and metabolic syndrome in normal glucose tolerance and normal fasting glucose subjects." *Acta Diabetologica* 48:79-88.

[9] Lai HM, Chen CJ, Su BY, Chen YC, Yu SF, Yen JH, Hsieh MC, Cheng TT, and Chang SJ. 2012. "Gout and type II Diabetes have incidences." *Rheumatology (Oxford)* 51:715-720.

[10] Shichiri, M, Iwamoto H, and Marumo F. 1990. "Diabetic hyperuricemia as an indicator of clinical nephropathy." *Am J Nephrol*. 10:115-22.

[11] Dehghan A, Van Hoek M, Sijbrands JG, Holfman A, and Witterman JCM. "High levels of serum Uric acid as a novel risk factor for type II Diabetes Mellitus." *Diabetes Care* 31: 361-62.

[12] Hayden MR, and Tyagi SC. "Uric acid: A new at an old risk marker for cardiovascular disease, metabolic syndrome and type II Diabetes Mellitus: The urate redox shuttle." *Nutr and Metab (Lond)* 1:10.

[13] Nakagawa T, Hu H, Zharikov S, and Finch JL, et al., 2005. "Oxypurinol improves coronary and peripheral endothelial function in patients with coronary artery disease." *Free Radic Biol Med* 39:1184-1190.

[14] Baldus S, Koster R, and Chumley P, et al., 2005. "Oxypurinol improves coronary and peripheral endothelial function in patients with coronary artery disease." *Free Radic Biol Med* 39: 1184-1190.

[15] Khosla UM, Zharikov S, and Finch JL, et al., "Hyperuricemia induce endothelial dysfunction." *Kidney Int* 67: 1739-1742.

[16] Roy D, Perreault M, and Marette A. 1998. "Insulin stimulation of glucose uptake in skeletal muscles and adipose tissues *in vivo* is NO-dependent." *Am J Physiol* 274: E692-699.

[17] Robles Cervantes JA, Ramos-Zavala MG, Gonzalez-Ortiz M, et al., 2011. "Relationship between serum concentration of uric acid and insulin secretion among adults with type II diabetes mellitus." *Int J Endocrinol*: 107904. http://dx.doi.org/10.1155/2011/107904

[18] Yoo TW, Sung KC, and Shin HS, et al., 2005. "Relationship between serum uric acid concentration and insulin resistance and metabolic syndrome." *Circ J* 69:928-933.

[19] Tsouli SG, Liberopoulos EN, and Mikhailidis DP et al., 2006. "Elevated serum Uric acid is associated with impaired glomerular filtration rate in metabolic syndrome: an active component or an innocent bystander"? *Metabolism* 55:1293-1301.

[20] Rosolowsky ET, Ficociello LH, and Maseli NJ et al., "High-normal serum uric acid in association with impaired glomerular filtration

rate in nonproteinuric patients with type 1 diabetes." *Clin J Am Soc Nephrol* 3:706-713.

[21] Qiao Q, Gao W, and Zhang L, et al., 2007. "Metabolic syndrome and cardiovascular disease." *AnnCli Biochem* 44: 232.

[22] Gogia A, and Agarwal PK. 2006. "Metabolic syndrome." *Indian J Med Sci* 60:72.

[23] American Heart Association: *Metabolic syndrome.* www.ameriacanheart.org/presenter.jhtml?identifier=4756. 2007.

[24] Thompson D, Pepys M, Wood S. 1999. "The physiological structure of human C- reactive protein and its complex with phosphocholine." *Structure* 7: 169-177. doi:10.1016/S0969-2126(99)80023-9.

[25] Lau D, Dhillon B, Yan H, Szmitko P, and Verma S. 2005. "Adipokines: a molecular link between obesity and atherosclerosis." *American Journal of physiology. Heart and Circulatory physiology* 288: H2031-41. doi: 10.1152/ajpheart. 01058.2004. PMID 15653761.

[26] Pepys M, and Hischfield G. 2003. "C-reactive protein: A Critical update." The *J of Clinical Investigation* 111: 1805-12. doi:10.1172/JCI18921.

[27] Guiseppe S, and Matteo PC. 2006. "Reactive protein in hypertension: clinical significance and predictive value." *Nutrition, Metabolism & Cardiovascular Diseases* 16:500-508.

[28] Kailash P. 2003. "C-Reactive protein and cardiovascular diseases." *Int J of Angiology* 12: 1-12.

[29] Pradhan A, J Manson, N Rifai, J Buring, and P Ridker. 2001. "C-reactive protein, interleukin-6, and risk of developing type II Diabetes Mellitus." *JAMA* 286: 327-334.

[30] Vahdat K, Jafari SM, Pazoki R, and Nabipour I. 2007. "Concurrent increased high sensitivity C-reactive protein and chronic infections are associated with coronary artery diseases: a population-based study." *Indian J. Med. Sci.* 61:135-143.

[31] Braham D, and Trident P. 1972. "An improved color reagent for the determination of blood glucose by the oxidase system." *Analyst* 40:1232-7.

[32] Han TS, Sattar N, William K, Gozales VC, Lean ME, and Haffiner SM. 2002. "A prospective study of C-reactive protein in relation to the development of diabetes and metabolic syndrome in Mexico city Diabetes Study." *Diabetes Care* 25:201621.

[33] Andersen HC. 1950. McCarthy M. *Am. J. Med* 8:445

In: Medical Bioinformatics and Biochemistry   ISBN: 978-1-53614-952-4
Editors: Rajneesh Prajapat et al.   © 2019 Nova Science Publishers, Inc.

Chapter 10

NANO-DIABETOLOGY

Rupesh Kumar Basniwal* and V. K. Jain
Amity Institute of Advanced Research and Studies
Amity University, Uttar Pradesh, India

ABSTRACT

Nanotechnology is the application of nanomaterials in various fields like agricultural, industrial, health, medicine etc. Nanomedicine is growing rapidly in health sector and creating new opportunities for medical science in disease treatment and management. This technology is further extended for treatment and management of the diabetes. The conventional treatment of diabetes is painful because patients should take regularly insulin injection according to blood glucose level. This painful situation and other associated complications of diabetes can overcome by advances in nanomedicine, like Glucose nano sensors, layer-bilayer (LBL) technique, Carbon Nanotubes and Quantum Dots (QD's) etc. Developing nano-oral insulin is also good alternate for the treatment of the disease. Late wound healing is also another problem of diabetic patient which can overcome by applications of nanoparticles (Al_2O_3, CeO_2, Y_2O_3, AuNPs). Nanotechnology in diabetes research has enabled the development of novel

---

* Corresponding Author's Email: rup4kumar@gmail.com.

glucose measurement and insulin delivery modalities, which hold the potential to dramatically improve quality of life for diabetic patient.

Keywords: nanotechnology, quantum dots, nanoparticles

INTRODUCTION

Diabetes is one of largest public health problem spreading at alarming worldwide [1, 2]. It's almost affecting 25.8 million people in the United States and 382 million worldwide [3]. In last few decades, the status of diabetes has been changed from a mild disorder to major cause of morbidity and mortality in the youth as well as in middle aged people [4]. Diabetes is the sixth leading cause of death in U.S and mostly affected by Type 2 diabetes [5, 6].

Conventional way to treat diabetes is daily subcutaneous insulin injections and pricking of finger for the measurement of blood glucose level [7]. Taking insulin injection on daily basis is painful process and which lead to patient noncompliance and result into overdoses of insulin [8]. This problem could overcome from such problem by application of nanotechnology.

Nanotechnology is a branch of science, in which characterization, production and behavior of nanoparticles is studied for various field applications like medical, industrial, agricultural etc. Sizes of nanoparticles can vary between 1to 100nm.Different physical, chemical and biological properties of nanoparticles compare to their bulk material makes them useful tool for biomedical applications [8, 9].

Nanoparticles applications, in diagnosis and in delivery of molecules (DNA, RNA and proteins) and in monitoring the progression of disease are already in trend for therapeutic purposes [10]. The biomedical applications are possible due to various formulations of nanoparticles like liposomes, polymer nanoparticles, nanostructures, metallic nanoparticles, stimuli-responsive nanoparticles and nanofabricated devices [11, 12, 13, 14, 15, 16, 17].

The size of nanoparticles matches with the size of biological material like a quantum dot is about the same size as a small protein (<10 nm) and drug-carrying nanostructures are the same size as some viruses (<100 nm). This similarity makes nanoparticles special and compatible tool for various sensor technology applications [6]. For example, as compare to conventional glucose sensor technology, the improved glucose sensor technology based on Nano-particles gives more accurate, frequent and convenient results for blood glucose estimations [18, 19, 20, 21]. This improved sensing technology will lead to more accurate measurement of insulin doses and further which helps in management of diabetes disease in a greater way.

Acute metabolic disturbances or chronic tissue damage are observed during severe case of diabetes [22, 23]. Accessibility of insulin was start from 1921 for controlling diabetes with medications. Gestational diabetes usually short-out after delivery but deficiency of proper treatments can cause many complications [24].

10.1. DETECTION OF INSULIN AND BLOOD SUGAR

The prefix of nanotechnology derives from '*nanos*' - the Greek word for dwarf [25]. The size of nanoparticles varies from 1 to 100nanometer. Nano material shows different physicochemical properties at nanoscale compare to their bulk nature. The development of nanotechnology at high pace rate but development of nanomedicine is still in neonatal stage. The nano medicines have a great potential to cure, treat and manage the diabetes disease. This is possible because nanoparticles have eligibility to cross the epithelial barriers in body due to their small size and which is challenging for conventional drugs. Just because of this nature they can be used as a drug carrier in form of nano-boat for enhancing drug delivery

 Rupesh Kumar Basniwal and V. K. Jain

and to treat diabetes disease. Drawbacks of conventional drug delivery system *i.e.* lacking in target specificity, altering effects and diminishing potency due to drug metabolism in the body [26] can be overcome by applications of nanoparticles and nanotechnology. The glucose sensors based on nanotechnology are playing crucial roles in monitoring of glucose level and in management of diabetes disease [27]. Measurement and controlling of insulin and blood sugar can be achieved in the following ways through the applications of nanotechnology:

10.1.1. By Micro-Physiometer

Construction of microphysiometer based on multi wall carbon nanotubes is done into very small tubes through stacking and rolling of carbon atoms. These are generating electrical signaling presence of small molecules of insulin or glucose and able to work on cellular pH level. It is useful for measuring chemical substance specially ligands for specific plasma membrane receptors [28]. Real time monitoring of insulin can be done with the help of this developed sensor. In this, insulin molecules oxidize in presence of glucose and generate current by transferring electrons to electrode and this continue till the last molecule of insulin oxidize but supply of glucose should be not limited.

10.1.2. By Implantable Sensor

Implantable sensor is capable in long-term monitoring of tissue glucose concentrations by wireless telemetry [29]. It is very helpful in controlling the glucose level of blood through measuring their concentration. In this method polyethylene glycol beads coated with fluorescent molecules are injected under the skin and stay in the interstitial fluid. Glucose displaces the fluorescent molecules and creating a glow when the interstitial fluid glucose drops to dangerous levels and this alarming situation is detected by sensor microchip [30]. Chip can be implanted under the skin for measuring various body parameters along with glucose i.e., pulse rate, temperature and pH level. Electrical signal generated from this sensor can be measure and monitor on regular basis [31].

10.2. METHODOLOGY

10.2.1. Conventional

Conventional way to treat diabetes was taking insulin doses at different interval inside the body through the injection mode which generates painful and stressful conditions for the patient. Taking insulin orally and releasing it into the bloodstream according to the need of patient can be alternate way to treat this disease.

The oral route is best way for drug administration because subcutaneous injection of insulin is painful process. The oral route administration of insulin can't be limited by their quantity because of hefty production of pharmaceutically active insulin is going on [32]. Generally, hydrophilic drugs do not diffuse across the bi-lipid layer wall of intestinal epithelium layer which limit their use in health sector. This bottleneck problem can be resolve with the help of nanotechnology. Intestinal permeation enhancers like chitosan can rectify this problem but not in effective way because of limited absorption of the drug. In vivo studies proved that absorption and retention of insulin can be further improved by application of nanoparticle with muco-adhesive chitosan. The drawback of this oral method was that insulin is sensitive to pH, temperature and digestive enzymes in stomach. To overcome from this drawback microsphere was developed.

10.2.2. Microsphere

Insulin encapsulated with microsphere act as protease inhibitors and permeation enhancers which supports the oral administration theory of insulin [33].

10.2.3. Nano-Pump

This is also another noninvasive method for controlling the diabetes. In this method nano pump with glucose sensor and insulin reservoir is fitted inside the body. This powerful pump can deliver insulin at constant rate in body and maintain the desire level of sugar in blood.

10.2.4. Artificial Pancreas

This automated artificial pancreas system is design to pump out insulin dose from insulin reservoirs as soon as change in blood glucose concentration and the dose of insulin is decided based on computer algorithm [34].

10.3. ROLE OF NANO TECHNOLOGY IN GLUCOSE MONITORING

The conventional finger-prick technique is painful process for measuring blood glucose level in body [35]. Micro-dialysis probes are also available in the market for glucose measurement, but their impaired response and unpredictable signal drift make them less reliable technique [36, 37, 38]. To overcome from aforesaid problem, application of nanotechnology is in trend.

10.3.1. Glucose Nanosensors

These are based on smart sensor technology like 'smart tattoo'. Fluorescence-based Nanosensors implanted into the skin for sensing glucose level. Lectins [39. 40], enzymes (hexokinase) [41] bacterial binding proteins [42, 43, 44, 45] and boronic acid derivatives [46] are various biological or artificial receptors which might be engineered for nano-sensors for measuring glucose concentration.

10.3.2. Layer-by-Layer (LBL) Technique

In this technology, a six bilayers thin film (approximately 10 nm) is prepared by electrostatically layer-by-layer (LBL) technique. Nano-assembly, alternating layers of positive and negative charge polymer is tunable, permeable and biocompatible. These nano films can be turn into micro-vesicles for glucose sensing purpose. This technique is noninvasive if based on fluorescence for sensing of blood glucose [47, 48, 49, 50, 51, 52].

10.3.3. Carbon Nanotubes

A micro-physiometer, a small device is developed by arranging multiwalled carbon atoms sheet in stack form and then rolled into small tube form. The chamber of this tube is used for measuring insulin concentration which is directly proportional to generation of current between the electrode and nanotube. Continuous measurement of insulin is also possible by measuring current continuously. This gives a real time measurement and monitoring of insulin concentration.

10.3.4. Quantum Dots (QDs)

Quantum Dots (2-10 nm) [53] is used as a fluorescent probe for biosensor applications in Fluorescence-resonance energy transfer (FRET). For example, glucose sensor based on FRET has QDs as a fluorescence donor and gold nanoparticles as an acceptor. The glucose displaces concanavalin A-labeled QDs from gold-labeled cyclodextrin, thereby, reducing FRET and increasing fluorescence [54].

10.4. DEVELOPMENT OF PATIENT-FRIENDLY INSULIN DELIVERY NANOPARTICLE FORMULATIONS

As compare to conventional drugs, nano-formulation of drugs has capacity to cross the epithelial barrier. Nanoparticles with bound ligands or with hydrophilic group have capacity to cross the epithelial barriers through trans-cytosis or para-cellular diffusion process. Recently antigen sampling micro-fold cells (M cells) based nano-formulation has been developed to cross the epithelial barriers. Currently, many methods like regeneration of β-cells, reprogramming of native cells and transplantation of insulin-producing cells are available to restore insulin production for the treatment of diabetes [55].

These methods are associated with their own limitations and challenges which can overcome by applications of nano technology. Protection of transplanted cells can also be done with the help of nanotechnology while suppling of oxygen, glucose, insulin and other necessary nutrients are not interrupted [56]. Protecting islet activity without interference in their function can also achieved by layer-by-layer polymer deposition [57] polyion complex formation [58] and chemical reactions of polymers methods [59].

Another application of nanotechnology is safe drug delivery and protection from unfavorable conditions like change in pH, temp or highly acidic environment. Delivery of gene encoding glucagon-like peptide-1 via nanoparticles to boost insulin secretion and islet viability [60, 61] is also in trend. Vaccine development based on nanoparticles is also in the process to prevent the autoimmune destruction of β-cells in patients with type 1 diabetes [62, 63]. Self-replicating and RNA-based antigens [64, 65] deliveries are also possible with application of nanotechnology for the treatment of diabetes.

10.4.1. Anti-Oxidative Role of Nanoparticles

Oxidative stress plays a major role in etiology of several diabetic complications [66; 67; 68]. Several nanoparticles work like free radical scavenger due to their variable oxidation state likeCeO_2, Y_2O_3 and Al_2O_3 (alumina) [69]. Nontoxic groups of nanoparticles can protect neutrophils and macrophages cells from oxidative stress and their antimicrobial and anti-oxidative nature makes them perfect agent for wound healing in diabetes [70; 71; 72; 73]. Other nanoparticles like gold nanoparticles and Ag+ loaded zirconium phosphate nanoparticles have crucial role in diabetic wound healing [74] Anti-oxidative [75] and antihyperglycemic activities of gold nanoparticles [76].

10.5. APPLICATIONS OF NANOMEDICINE IN HEALTH CARE AND DISEASE

Sky rocket growth of nanotechnology and engineered nanomaterials is due to their huge potential applications in different sectors. These nanoparticles are playing crucial role in the field of medicine. Ability to cross the blood brain barrier makes them useful for drug delivery and in treatment of neurodegenerative disorder. Generally, nanoparticles due to large surface area and their uniqueness nature shows hyper and different useful activity compare to their bulk material nature. To avoid toxic effect on human body their application should be in proper and optimum concentration. Continuous growth of Nano-medicines is compulsory for social, economic and health progress benefits. Nano-diagnostics, nano-pharmaceuticals, reconstructive surgery, nanorobotics, nano-surgery, regenerative medicine and ultrafast DNA sequencing are various current fields of nanotechnology [77, 78].

CONCLUSION

Management of diabetes based on nanotechnology is new but rapidly growing due to successful in various disease treatments. The first nanoparticle-based therapy for cancer treatment received FDA approval in 1995 [79, 80, 81, 82]. In this a pegylated liposome nanoparticle formulation loaded with the chemotherapeutic agent doxorubicin was used. Since, 20 different nano-medicines formulations are currently under clinical trial for cancer treatment. Cardiovascular disease [83, 84] is also manageable by application of nano-particles like; enhancement of MRI contrast image for the monitoring of acute myocardial infraction in human patients [85, 86].

As further applications of these nanoparticles can be extending towards management of diabetes disease. Immune modulation for cell-based therapies, non-invasive monitoring of disease progression, continuous monitoring of blood glucose levels and development of patient-friendly insulin are the areas for controlling diabetes disease based on nanoparticles. The next generation sensors are nanosensors and their sensitivity and specificity are far better than conventional sensors.

Nano-sensor based technology can be further utilized for insulin delivery in diabetic patient because their results are more consistent and reliable, with less drift from sensor degradation or failure. Reducing the immune response towards new insulin-producing cells is also new area in cell-based therapy for nanotechnology. This can be achieved through therapeutic targeted delivery of nucleic acid to minimize the immune response against newly insulin-producing cells [87]. This can play major role in transplantation technology [88]. Applications of nanoparticles is further limited by bottleneck problem i.e., toxicity at higher concentration. Recently, FDA has issued guidelines for fostering the safe development of nanotechnology-based products for clinical use. Safety and long-term performance must be fully evaluated in the design of diabetes therapeutics and diagnostics, especially for materials that are not degraded or cleared from the body.

In summary, nanotechnology has great role in improving the management of diabetes. FDA-based nanotechnology formulations are in demand for controlling diabetes and encouraging noninvasive methods like air route i.e., pulmonary rote for painless treatment of the disease. Now the time is demanding for robust glucose-sensitive nanoparticles and their development for nanodevices and the improvement of integrated glucose-sensing and insulin-delivering nano formulations.

REFERENCES

[1] Huizinga MM, and Rothman RL. 2006. "Addressing the diabetes pandemic: A comprehensive approach." *Indian J Med Res* 124:481-484.

[2] Wild S, Roglic G, Green A, Sicree R, and King H. 2004. "Global prevalence of diabetes: Estimates for the year 2000 and projections for 2030." *Diabetes Care* 27: 1047-53.

[3] Bratlie KM, York RL, Invernale MA, Langer R, and Anderson DG. 2012. "Materials for diabetes therapeutics." *Advanced Healthcare Materials* 1: 267–284.

[4] Zhang P, Zhang X, Brown J, Vistisen D, Sicree R, Shaw J, and Nichols G. 2010. "Global healthcare expenditure on diabetes for

2010 and 2030." *Diabetes Res Clin Pract.* 87: 293–301. doi: 10.1016/j.diabres.2010.01.026.

[5] Ross SA, Gulve EA, Wang M. 2004. "Chemistry and Biochemistry of Type 2 Diabetes." *Chemical Reviews* 104:1255–1282.

[6] Sowers JR, and Lester MA. 1999. "Diabetes and cardiovascular disease." *Diabetes care* 22:C14–20.

[7] Mo R, Jiang T, Di J, Tai W, and Gu Z.2014. "Emerging micro- and nanotechnology based synthetic approaches for insulin delivery." *Chemical Society Reviews.* 43: 3595–3629.

[8] Whitesides GM. 2003. "The 'right' size in nanobiotechnology." *Nature Biotech.* 21: 1161–1165.

[9] LaVan DA, Lynn DM, and Langer R. 2002. "Moving smaller in drug discovery and delivery." *Nature Rev. Drug Discov.* 1: 77–84.

[10] McNeil SE. 2011. "Unique benefits of nanotechnology to drug delivery and diagnostics." *Methods Mol. Biol.* 697, 3–8.

[11] Venkatraman, S S, Ma, L L, Natarajan, J V & Chattopadhyay, S. Polymer- and liposome-based nanoparticles in targeted drug delivery. *Front. Biosci.*2, 801–814 (2010).

[12] Stinchcombe TE. 2007. "Nanoparticle albumin-bound paclitaxel: a novel Cremphor-EL-free formulation of paclitaxel." *Nanomed.* 2: 415–423.

[13] Barbas AS, Mi J, Clary BM, and White RR. 2010. "Aptamer applications for targeted cancer therapy." *Future Oncol.* 6: 1117–1126.

[14] Chiu GN, et al., 2009. "Lipid-based nanoparticulate systems for the delivery of anti-cancer drug cocktails: Implications on pharmacokinetics and drug toxicities." *Curr. Drug Metab.* 10: 861–874.

[15] Schroeder A, Levins CG, Cortez C, Langer R, and Anderson DG. 2010. "Lipid-based nanotherapeutics for siRNA delivery." *J. Intern. Med.* 267: 9–21.

[16] Veiseh O, Gunn JW, and Zhang M. 2010. "Design and fabrication of magnetic nanoparticles for targeted drug delivery and imaging." *Adv. Drug Deliv. Rev.* 62: 284–304.

[17] Leonard F and Talin AA. 2011. "Electrical contacts to one-and two-dimensional nanomaterials." *Nature Nanotech.* 6: 773–783.

[18] Chun AL. 2006. "Nanosensors: Bring it on." *Nature Nanotechnology* 84.

[19] Bahshi L, Freeman R, Gill R, and Willner I. 2009. "Optical Detection of Glucose by Means of Metal Nanoparticles or Semiconductor Quantum Dots." *Small* 5: 676–680.

[20] Veetil JV, Jin S, and Ye K. 2010. "A glucose sensor protein for continuous glucose monitoring." *Biosensors and Bioelectronics* 26: 1650–1655.

[21] Gordijo CR, Shuhendler AJ, and Wu XY. 2010. "Glucose-Responsive Bioinorganic Nanohybrid Membrane for Self-Regulated Insulin Release." *Adv Functional Materials* 20:1404–1412.

[22] Rother KI. 2007. "Diabetes treatment - bridging the divide." *The New England J of Medicine* 356: 1499–501.

[23] Lawrence JM, Contreras R, ct al., 2008. "Trends in the prevalence of preexisting diabetes and gestational diabetes mellitus among a racially/ethnically diverse population of pregnant women, 1999–2005." *Diabetes Care* 31: 899–904.

[24] Lambert P. "What is Type 1 Diabetes?." Medicine 30: 2002; 1–5.

[25] Rathod B Kinjal et al., 2011. "Glimpses of current advances of nanotechnology in therapeutics." *Int J of Pharmacy and Pharmaceutical Sci* 3: 8-12

[26] Subramani K. 2006. "Applications of nanotechnology in drug delivery systems for the treatment of cancer and diabetes. *Int J of Nanotechnology* 3: 557 - 580.

[27] Samuel D, Bharali D. 2011."The role of nanotechnology in diabetes treatment: current and future perspectives." *Int J of Nanotechnology* 8:53 - 65.

[28] Connell Mc HM, et al., 1992. "The cytosensormicrophysiometer: biological applications of silicon technology." *Science Live Journal.* 257:1906-1912.

[29] Monda Erin. 2010. *Implantable glucose sensors.* Health tech zone medical featured article. 2: 2-6.

[30] Gough AD, et al., 2010. "Bioengineering and Diabetes Function of an Implanted Tissue Glucose Sensor for More than 1 Year in Animals." *Sci Translational Med J* 2: 42-53.

[31] Kumar Arya, et al., 2008. "Applications of nanotechnology in diabetes." *Digest J of Nanomaterials and Biostructures.* 3: 221-225.

[32] Gordon JS. 2002. "Development of oral insulin progress and currents status." *Diabetes Metab Res Rev.* 18: S29-37.

[33] Jindal KS. 2009. "Formulation and evaluation of insulin enteric microspheres for oral drug delivery." *Acta Pharmaceutica Sciencia* 51: 121-127.

[34] Crane Mark. 2011. "FDA issues graft guidance for artificial pancreas." *Medscape Today.* Vol 1.

[35] Moghimi SM, Hunter AC, and Murray JC. 2005. "Nanomedicine: current status and future prospects." *FASEB J* 19: 311-330.

[36] Mastrototaro JJ. 2000. "The MiniMed continuous glucose monitoring system." *Diabetes Technol Ther* 2: S13-S18.

[37] Maran A, Crepaldi C, Tiengo A, Grassi G, and Vitali E, et al., 2002. "Continuous subcutaneous glucose monitoring in diabetic patients: A multicentre analysis." *Diabetes Care* 25: 347-352.

[38] Garg S, Zisser H, Schwarz S, Bailey T, and Kaplan R, et al., 2006. "Improvement in glycemic excursions with a transcutaneous, real-time continuous glucose sensor: a randomized controlled trial." *Diabetes Care* 29: 44-50.

[39] McCartney LJ, Pickup JC, Rolinski OJ, and Birch DJ. 2001. "Near-infrared fluorescence lifetime assay for serum glucose based on allo phyco cyaninl abeledconcana valin A. *Anal Biochem* 292: 216-221.

[40] Ballerstadt R, and Schulz JS. 2000. "A fluorescence affinity hollow fiber sensor for continuous transdermal glucose monitoring." *Anal Chem* 72: 4185-4192.

[41] Hussain F, Birch DJS, and Pickup JC. 2005. "Glucose sensing based on the intrinsic fluorescence of sol-gel immobilized yeast hexokinase." *Anal Biochem* 339: 137-143.

[42] Marvin JS, and Hellinga HW. 1998. "Engineering biosensors by introducing fluorescence allosteric signal transduction: Construction of a novel glucose sensor." *J Am Chem Soc* 120: 7-11.

[43] Tolosa L, Gryczynski I, Eichhorn LR, Dattelbaum JD, and Castellano FN, et al., 1999. "Glucose sensor for low-cost lifetime-based sensing using a genetically engineered protein." *Anal Biochem* 267: 114-120.

[44] Salins LL, Ware RA, Ensor CM, and Daunert S. 2001. "A novel reagentless sensing system for measuring glucose based on the galactose/glucose-binding protein." *Anal Biochem* 294: 19-26.

[45] Ye K, and Schultz JS. 2003. "Genetic engineering of an allosterically based glucose indicator protein for continuous glucose monitoring by fluorescence resonance energy transfer." *Anal Chem* 75: 3451-3459.

[46] Pickup JC, Hussain F, Evans ND, Rolinski OJ, and Birch DJ. 2005. "Fluorescencebased glucose sensors." *Biosens Bioelectron* 20: 2555-2565.

[47] Ariga K, Hill JP, and Ji Q. 2007. "Layer-by-layer assembly as a versatile bottomup nanofabrication technique for exploratory research and realistic application." *Phys Chem Chem Phys* 9: 2319-2340.

[48] Trau D, and Rennenberg R. 2003. "Encapsulation of glucose oxidase microparticles within a nanoscale layer-by-layer film: immobilization and biosensor applications." *Biosens Bioelectron* 18: 1491-1499.

[49] Chinnayelka S, and McShane MJ. 2006. "Glucose sensors based on microcapsules containing an orange/red competitive binding

resonance energy transfer assay." *Diabetes Technol Ther* 8: 269-278.

[50] McShane M, and Ritter D. 2010. "Microcapsules as optical biosensors." *J Mater Chem* 20: 8189-8193.

[51] Zhi ZL, and Haynie DT. 2006. "High-capacity functional protein encapsulation in nanoengineered polypeptide microcapsules." *Chem Commun* 14: 147- 149.

[52] Pickup JC, Zhi ZL, Khan F, Saxl T, and Birch DJ. 2008. "Nanomedicine and its potential in diabetes research and practice." *Diabetes Metab Res Rev* 24: 604- 610.

[53] Michalet X, Pinaud FF, Bentolila LA, Tsav JM, and Doose S, et al., 2005. "Quantum dots for live cells, *in vivo* imaging, and diagnostics." *Science* 307: 538-544.

[54] Tang B, Cao L, Xu K, Zhuo L, and Ge J, et al., 2008. "A new biosensor for glucose with high sensitivity and selectivity in serum based on fluorescence resonance energy transfer (FRET) between CdTe quantum dots and Au nanoparticles." *Chemistry* 14: 3637-3644.

[55] Robertson RP. 2004. "Islet transplantation as a treatment for diabetes , a work in progress."*N. Engl. J. Med.* 350: 694–705.

[56] Lanza RP, Hayes JL, and Chick WL. 1996. "Encapsulated cell technology." *Nature Biotech.* 14: 1107–1111.

[57] Wilson JT, Cui W, and Chaikof EL. 2008. "Layer-by-layer assembly of a conformal nanothin PEG coating for intraportal islet transplantation." *Nano Lett.* 8: 1940–1948.

[58] Krol S, et al, 2006. "Multilayer nanoencapsulation.New approach for immune protection of human pancreatic islets." *Nano Lett.* 6: 1933–1939.

[59] Contreras JL, et al., 2004. "A novel approach to xenotransplantation combining surface engineering and genetic modification of isolated adult porcine islets." *Surgery* 136: 537–547.

[60] Li F, and Mahato RI. 2010. "RNA interference for improving the outcome of islet transplantation." *Adv. Drug Deliv. Rev.* 63:47–68.

[61] Oh S, Lee M, Ko KS, Choi S, and Kim SW. 2003. "GLP-1 gene delivery for the treatment of type 2 diabetes." *Mol. Ther.* 7: 478–483.

[62] Moon JJ, Huang B, and Irvine DJ. 2012. "Engineering nano- and microparticles to tune immunity." *Adv. Mater.* 24: 3724–3746.

[63] Nembrini C, et al., 2011. "Nanoparticle conjugation of antigen enhances cytotoxic T-cell responses in pulmonary vaccination." *Proc. Natl Acad. Sci.* 108: E989–E997.

[64] Geall AJ, et al., 2012. "Nonviral delivery of self-amplifying RNA vaccines." *Proc. Natl Acad. Sci.* 109: 14604–14609.

[65] Nguyen DN, et al., 2012. "Lipid-derived nanoparticles for immunostimulatory RNA adjuvant delivery." *Proc. Natl Acad. Sci.* 109: E797–E803.

[66] Manish M, Hemant K, and Kamlakar T. 2008. "Diabetic delayed wound healing and the role of silvernanoparticles." *Digest J of Nanomaterials and Biostructures* 3: 49-54.

[67] Giugliano D, Ceriello A, and Paolisso G. 1996. "Oxidative stress and diabetic vascular complications." *Diabetes Care* 19: 257-267.

[68] Feldman EL, Stevens MJ, and Greene DA. 1997. "Pathogenesis of diabetic neuropathy." *Clin Neurosci* 4: 365-370.

[69] Kingery WD, Bowen HK, and Uhlman DR. 1976. *Introduction to Ceramics.* John Wiley, New York.

[70] Wright JB, Lam K, Hanson D, and Burrell RE. 1999. "Efficacy of topical silver against fungal burn wound pathogens." *Am J Inf Cont* 27: 344-350.

[71] Demling RH, and DeSanti L. 2006. "Effects of silver on wound management." *Wounds* 13: 4.

[72] Monafo WW, and Moyer CA. 1968. "The treatment of extensive thermal burns with 0.5% silver nitrate solution." *Ann NY Acad Sci* 150: 937-945.

[73] Fox CL Jr. 1968. "Silver sulfadiazine--a new topical therapy for Pseudomonas in burns.Therapy of Pseudomonas infection in burns." *Arch Surg* 96: 184-188.

[74] Yin HQ, Langford R, and Burrell RE. 1999. "Comparative evaluation of the antimicrobial activity of ACTICOAT antimicrobial barrier dressing." *J Bum Care Rehab* 20: 195-200.

[75] Ruggiero D, Lecomte M, Michoud E, Lagarde M, and Wiernsperger N. 1997. "Involvement of cell-cell interactions in the pathogenesis of diabetic retinopathy." *Diabetes Metab* 23: 30-42.

[76] BarathManiKanth S, Kalishwaralal K, Sriram M, Pandian SR, and Youn HS, et al., 2010. "Anti-oxidant effect of gold nanoparticles restrains hyperglycemicconditions in diabetic mice." *J Nanobiotechnology* 8: 16.

[77] Jain KK. 2008. "Nanomedicine: application of nanobiotechnology in medical practice." *Med Princ Pract* 17:89-101.

[78] Surendiran A, Sandhiya S, Pradhan SC, and Adithan C. 2009. "Novel applications of nanotechnology in medicine." *Indian J Med Res* 130: 689-701.

[79] Hrkach J, et al., 2012. "Preclinical development and clinical translation of a PSMA-targeted docetaxel nanoparticle with a differentiated pharmacological profile." *Sci. Transl. Med.* 4: 128-39.

[80] Davis ME, et al., 2010. "Evidence of RNAi in humans from systemically administered siRNA via targeted nanoparticles." *Nature* 464: 1067–1070.

[81] Peer D, et al., 2007. "Nanocarriers as an emerging platform for cancer therapy." *Nature Nanotech.* 2: 751–760.

[82] Barenholz Y, and Doxil R. 2012. "The first FDA-approved nanodrug: lessons learned." *J. Control Release* 160: 117–134.

[83] McCarthy JR. 2010. "Nanomedicine and cardiovascular disease." *Curr.Cardiovasc.Imag.Rep.* 3: 42–49.

[84] Lobatto ME, FusterV, Fayad ZA, and Mulder WJ. 2011. "Perspectives and opportunities for nanomedicine in the

management of atherosclerosis." *Nature Rev. Drug Discov.* 10: 835–852.

[85] Alam SR, et al., "Ultrasmall superparamagnetic particles of iron oxide in patients with acute myocardial infarction: early clinical experience." *Circ. Cardiovasc. Imag.* 5: 559-565.

[86] Kim BYS, Rutka JT, and Chan WCW. 2010. "Nanomedicine." *New Engl. J. Med.* 363: 2434–2443.

[87] Whitehead KA, Langer R, and Anderson DG. 2009. "Knocking down barriers: advances in siRNA delivery." *Nature Rev. Drug Discov.* 8: 129-138.

[88] Yi P, Park JS, and Melton DA. 2013. "Betatrophin: a hormone that controls pancreatic β-cell proliferation." *Cell* 153: 747–758.

In: Medical Bioinformatics and Biochemistry   ISBN: 978-1-53614-952-4
Editors: Rajneesh Prajapat et al.   © 2019 Nova Science Publishers, Inc.

Chapter 11

EFFECT OF AJOWAN SEED POWDER ON OXIDATIVE STRESS IN NORMAL AND NIDDM PATIENTS

*Parul Gupta**
Department of Biochemistry, GSVM Medical College,
Kanpur, Uttar Pradesh, India

ABSTRACT

Diabetes mellitus is a group of metabolic disorders with one common manifestation: hyperglycaemia. The rationale behind this study is to see the effect of Ajowan seed powder on different oxidative stress parameters in blood/ serum. It was a hospital based prospective study conducted for a period of 8 months from September 2013- April 2014 in LLR hospital and diabetic clinic of GSVM medical college, Kanpur (UP). A total of 180 study participants were included in the study which consists of 90 diabetic patients and 90 healthy individuals. The investigation reports were collected from pathology laboratory of the college for both diabetic and normal subjects. The parameters which were studied includes catalase

---

* Corresponding Author's Email: parul2080@gmail.com.

(CAT), gluthatione peroxidise (GPx), glutathione reductase (GR), plasma Maloaldialdehyde (MDA), superoxide dismutase (SOD) before conducting the study prior consent of the subjects is taken. Epi-info software was used. And both paired and unpaired student 't' test were applied for analysis. Present study showed a significant change in plasma MDA level and increasing anti-oxidant activity which counteracts hyperglycemia when Ajowan powder was administrated. Observation suggest that Ajowan possess the anti-oxidant property.

Keywords: diabetes, hyperglycemia, Ajowan, oxidative stress

INTRODUCTION

Diabetes mellitus is a global public health problem of epidemic proportions, and its incidence is on the rise. Diabetes mellitus is a group of metabolic disorders with one common manifestation, hyperglycemia. It is characterized by insulin resistance in peripheral tissue and an insulin secretary defect of the β-cell [1, 2]. A currently favoured hypothesis is that oxidative stress, through a single unifying mechanism of superoxide production, is the common pathogenic factor leading to insulin resistance, β-cell dysfunction, impaired glucose tolerance (IGT) and ultimately to type 2 DM [3, 4].

The body's defence against oxidative stress is accomplished by interconnecting systems of antioxidant micronutrients (vitamins and minerals) and enzymes. In certain medicinal plants like Ajowan [5, 6] have been reported as effective hypolipidemic agent in normal and diabetes mellitus as well. It's effect on the level of oxidative stress, a positive causative factor for diabetes mellitus and on anti-oxidative defence enzyme system have not been studied thoroughly. Ajowan seed powder in modulating hyperglycemia induced oxidative stress may be responsible for alteration and complications in carbohydrate metabolism and possibly depressed antioxidant defence system in diabetes mellitus [7, 8]. Not many studies have been seen in India showing the effect of Ajowan seed powder on anti-oxidants and no studies were conducted in this region regarding this, Therefore, the rationale behind this study is to see the effect of ajowan seed powder on different and oxidative stress parameters in blood / serum and to correlate its effect on these parameters among diabetic patients [10].

11.1. METHODOLOGY

The methodology of this study is similar to the study conducted previously by the same author [REF]. It was a hospital based prospective study conducted for a period of 8 months from September 2013- April 2014 in LLR hospital and diabetic clinic of GSVM medical college, Kanpur (UP). The study participants were taken between the age group of >=18 to 60 years. A total of 180 study participants were included in the study which consists of 90 diabetic patients and 90 healthy individuals. The parameters which were studied are catalase (CAT), gluthatione peroxidase (GPx), glutathione reductase (GR), plasma maloaldialdehyde (MDA), superoxide dismutase (SOD). These parameters were measured with the suitable method available in the department of biochemistry of medical college. Selected subjects were advised to take Ajowan seed powder from last one month of the study to see its effect.

11.1.1. Statistical Analysis

Collected data were consolidated on excel sheets and further analyzed in Epi-info software. Mean along with standard deviation (SD) were calculated for the different parameters at different days of the study. Paired and unpaired student 't' test was employed to analyze the effect of drug between matched and unmatched groups. P value <0.05 is considered statistically significant.

                    Parul Gupta

11.2. RESULTS

Diabetes mellitus is a global public health problem of epidemic proportions, and its incidences on the rise [11]. In our study we showed the effect of Ajowan seed powder administration to see the effect on different parameters in both normal and NIDDM subjects.

Table 11.1. Age and sex wise distribution of study subjects during Ajowan seed powder administration (n = 180)

Age (years)	Normal subjects (n=90)		NIDDM subjects (n=90)	
	Male	Female	Male	Female
>=18-29	9	7	2	0
30-39	8	3	2	1
40-49	22	11	17	10
50-60	17	13	37	21
Total	56 (62.2)	34 (37.8)	58 (64)	32 (36)

*values within parenthesis are percentage.

Table 11.1 shows age and sex wise distribution of study subjects; the subjects were male preponderance 62% in normal while 64% in NIIDM. Among both the subjects the age groups between 50-60 years has maximum number of subjects. Followed by 40-49 years which itself shows either the diabetes was late diagnosed, or the subjects admitted themselves only after symptoms of diabetes.

Table 11.2 shows the effect of different parameters after continuous Ajowan seed powder administration in normal subjects. The Mean ± S.D, μmol/l MDA levels found to be at 1.44 ± 3.8 0th day, at 1.18 ± 1.64 30th day and 1.15 ± 1.326 on last month which showed a continuous trend of oxidative stress decline upon feeding of Ajowan seed powder to the normal subjects. This decline was found to be highly significant at different days. Observations suggest that Ajowan seed possess antioxidant property.

Table 11.2. Effect on values of studied parameters at different days of Ajowan seed powder administration and after withdrawal in Normal subjects

S No.	Parameter	0th day	30th day	Last month	15 days after withdrawal
1.	Maloaldaildehyde (MDA)	1.44 ± 3.8	1.183 ± 1.64 (2.7↓) p = 0.01	1.15 ± 1.83 (3.4↓) p = 0.02	1.29 ± 1.32 (6.4↑) p = 0.02
2.	Glutathione reductase (GR)	16.97 ± 1.43	17.75 ± 2 (4.6↑) p = 0.04	17.63 ± 1.96 (3.9↑) p = 0.02	16.97 ± 0.81 (3.8↓) p = 0.02
3.	Superoxide dismutase (SOD)	4.41 ± 0.31	4.9 ± 0.88 (11.1↑) p = 0.04	5.09 ± 1.1 (15.4↑) p = 0.03	4.83 ± 1 (5↓) p = 0.03
4.	Gluthatione peroxidase (GPx)	16.5 ± 1.17	17.27 ± 1.82 (4.2↑) p = 0.02	17.57 ± 1.94 (6.5↑) p = 0.01	16.87 ± 1.68 (4↓) p = 0.01
5.	Catalase (CAT)	144.87 ± 4.74	145.13 ± 7.4 (3↑) p = 0.01	150.23 ± 11.51 (5.1↑) p = 0.01	149.33 ± 8.22 (1.3↓) p > 0.05

The Mean ± S.D, Unit/gm Hb specific activity of GR was seen to be 16.97 ± 1.43 at 0th day, 17.75 ± 2.01 on 30th day and 17.63 ± 1.96 on last month. The enzyme activity increased significantly 4.6% on 30th day and 3.9% on last month from the enzyme activity before start of the drug. Observations suggest that Ajowan possess antioxidant property by increasing GR activity.

The Mean ± S.D, Unit/gm Hb specific activity of SOD was observed to be 4.41 ± 0.31 at 0th day, 4.9 ± 0.80 on 30th day and 5.09 ± 1.1 on last month. The enzyme activity increased significantly 11.1% on 30th day and 15.4% on last month from the activity before start of the drug. However, it decreased significantly by 5% on 15th day after drug withdrawal from last month. Observations suggest that Ajowan possess antioxidant property by upregulating SOD activity.

The Mean ± S.D, Unit/gm Hb specific activity of GPx was found to be 16.5 ± 1.17 at 0th day, 17.2 ± 1.81 on 30th day and 17.57 ± 1.94 on last month. Observations suggest that Ajowan possess antioxidant property by escalating GPx activity.

The Mean ± S.D, Unit/gm Hb specific activity of CAT was found to be 144.0 ± 4.18 at 0th day, 147.53 ± 7.96 on 30th day and 150.13±12.14 on last month. Observations suggest that Ajowan may possess antioxidant property by elevating CAT activity.

From the present study it was found that after administration of Ajowan seed powder for last months, the percentage of MDA level declined statistically significant by 3.4% in normal subjects and by 7.13% in diabetic subjects. Our results were like studies conducted by [12, 13] on streptozotocin diabetic animals.

It was found that Ajowan seed powder can increase level of glutathione reductase statistically significant by 3.9% and by 1.7% in normal and diabetic subjects. Studies were done previously to see these effects [14, 15]. It suggests that Ajowan has property of increasing GR activity level. Antioxidant enzymes activities (GPx and CAT) were observed to be upregulated statistically significant by Ajowan seed powder administration. The specific activity of GPx increased statistically significant by 4.8% in diabetics and by 6.5% in normal subjects. Similarly, the rise of CAT activity statistically significant was 5.1% in normal subjects and 4.1% in diabetic subjects. The results similar to studies conducted by Sathishsekar D [8], and Asli S, Alaattin S [9], showing significant increase in antioxidant enzymes like catalase and glutathione peroxidase on alloxan diabetic rats when treated ajowan seed powder. In the present study it was suggested that Ajowan possess anti-oxidant property by increasing SOD activity (16). These results are like the study conducted by Sathishsekar.

CONCLUSION

Based on our results the anti-oxidant property possessed by Ajowan seed powder is due to increasing, up- regulating, escalating the activities of these parameters like MDA, GR, SOD, GPx, CAT which helps in preventing hyperglycemia induced oxidative stress. It was found that MDA level was declined statistically significant by administration of Ajowan seed powder. Down regulation of antioxidant enzymes statistically significant was observed in diabetic patients. Antioxidant enzymes (GPx, GR, SOD and CAT) activities were observing to be intensified by Ajowan seed powder administration.

It suggests that Ajowan seed powder may possess some constituent, which has some antioxidant properties. Further studies are necessary to find out these constituents present in Ajowan seed powder.

REFERENCES

[1]   American Diabetes Association (ADA): Clinical Practice Recommendations. 2013. Report of the Expert Committee on the Diagnosis and Classification of Diabetes Mellitus. *Diabetes Care* 24 (1).

[2]   Diabetes Care, Report of the Expert Committee on the Diagnosis and Classification of Diabetes Mellitus. 1997. *Diabetes Care* 20: 1183-1197. doi.org/10.2337/diacare.20.7.1183.

[3]   International Diabetes Federation. *Diabetes e-Atlas*. 2005. http://www.eatlas.idf.org. 2005.

[4]   Gupta P, Srivastava AK. 2016. "Effect of Indigenous Drug Administration (KARELA) on Different Parameters Studied Among Healthy and NIDDM Subjects: A Hospital Based Study." *Int J of Integrative Med Sci* 3: 312-317 DOI: 10.16965/ ijims.2016.127.

[5]   Pandya N. 2007. Facts and Comparisons 4.0. *Journal of the Medical Library Association*, 95: 217–219. http://doi.org/10.3163/
1536-5050.95.2.217.

[6]   Booth GL, Kapral MK, Fung K, and Tu JV. 2006. "Relation between age and cardiovascular disease in men and women with diabetes compared with non-diabetic people: a population-based retrospective cohort study." *Lancet* 368:29-36.

[7]   Ahmed N, Hassan MR, Halder H, and Bennoor KS. 1999. Effect of Momordica charantia (Karolla) extracts on fasting and postprandial serum glucose level in NIDDM patients. *Bangladesh Med Res Counc Bull* 25:11-13.

[8]   Sathishsekar D, and Subramanian S. 2005. "Antioxidant properties of Momordica Charantia (bitter gourd) seeds on Streptozotocin induced diabetic rats. *Asia Pac J Clin Nutr.* 14:153-8.

[9]   Asli Semiz, and Alaattin Sen. 2007. "Antioxidant and chemoprotective properties of Momordica charantia L. (bitter melon) fruit extract." *African J of Biotechnology* 6: 273-277.

[10] Kumar MA, et al., 2001. *Biochemical studies on the status of free radical and scavangers of some of the bioactive natural products against urolothiasis.* PhD Thesis, University of Kanpur, 2001.

[11] Miura T, Itoh, C, Iwamoto N, Kato M, Kawai M, Park SR, and Suzuki I. 2001. "Hypoglycemic activity of the fruit of *Momordica charantia* in type 2 diabetic mice." *J. Nutr. Sci. Vitaminol. (Tokyo)* 47: 240–244.

[12] M Pawa. 2005. "Use of traditional/herbal remedies by Indo-Asian people with type 2 diabetes." *Pract Diab Int* 22:8.

[13] Tongia A, Tongia, SK, and Dave M. 2004. "Phytochemical determination and extraction of Momordica charantia fruit and its hypoglycaemic potentiation of oral hypoglycemic drugs in diabetes mellitus (NIDDM). *Indian J Physiol. Pharmacol.* 48, 241– 244.

[14] Shetty AK, Kumar GS, Sambaiah K, and Salimath PV. 2005. "Effect of bitter gourd (*Momordica charantia*) on glycemic status

in streptozotocin induced diabetic rats." *Plant Foods Hum. Nutr* 60: 109–112.

[15] Senanayake GV, Maruyama M, Sakono M, Fukuda N, Morishita T, Yukizaki C, Kawano M, and Ohta H. 2004. "The effects of bitter melon (Momordica charantia) extracts on serum and liver lipid parameters in hamsters fed cholesterol-free and cholesterol-enriched diets." *J Nutr Sci Vitaminol (Tokyo)* 50:253-7.

[16] Yadav UC, Moorthy K, and Baquer NZ. 2004. "Effects of sodium-orthovanadate and Trigonella foenum-graecum seeds on hepatic and renal lipogenic enzymes and lipid profile during alloxan diabetes." *J Biosci.* 29: 81-91.

About the Editors

Rajneesh Prajapat, PhD

Department of Medical Biochemistry,
Faculty of Medical Sciences, Rama Medical College
and Hospital (NCR), Rama University, Kanpur, UP, India

Dr. Rajneesh Prajapat recently completed his second MSc in Medical Biochemistry from Department of Med. Biochemistry, Rama Medical College - Hospital & Research Centre, Rama University, Kanpur, Uttar Pradesh, India, under the supervision of Dr. Ijen Bhattacharya. He completed his PhD from Department of Science, Mody University of

Science and Technology, Lakshmangarh, Sikar, Rajasthan, India under the supervision of Dr. R. K. Gaur on "Molecular and In Silico characterization of Begomovirus components infecting weeds of North India." He graduated from M. D. S. University, Ajmer, Rajasthan, India in 2003 and did MSc Biotechnology from University of Rajasthan, Jaipur, Rajasthan, India in year 2005. He has 45 international and 10 national publications and also 45 sequences in NCBI database. He is the founder of GVDB database and in silico Modelling studies on gemini virus and begomovirus. He has command over various molecular and bioinformatics tools and techniques. His core interests are homology modelling, molecular docking, biological database development and in silico drug designing. He worked as assistant professor at department of biotechnology, Sobhahasaria Group of Institution, Sikar, Rajasthan, India till January 2012.

Dr. Ijen Bhattacharya, MD
Department of Medical Biochemistry,
Faculty of Medical Sciences, Rama Medical College
and Hospital (NCR), Rama University, Kanpur, UP, India

Dr. Ijen Bhattacharya is presently working as Head and Professor, Department of Biochemistry, Rama Medical College - Hospital &

Research Centre, Rama University, Kanpur, Uttar Pradesh, India. He did his MD in Biochemistry from Guwahati Medical College (Guwahati University) Assam in the year 2005 and MBBS from Guwahati Medical College (Guwahati University) Assam in the year 2000. In 2011 he had also completed MSc (Clinical Research & Regulatory Affairs) from Sikkim Manipal University, INDIA. He worked as a Medical Officer in several nursing homes in Guwahati for a period of two years from Aug 2000 to July 2002, and as Assistant Professor at Kamineni Institute of Medical Science, Narketpally, A.P, India from 2005 to 2010. He has made significant contributions on Medical Biochemistry and published 30 national/international papers and presented more than 40 papers in national and international conferences. He has also visited foreign country for the sake of attending the conference/workshop. He is also a member of national and international medical societies. Presently, he is working on the biochemical analysis of various human diseases and bioinformatics characterization of diabetes associated genes. His core interests are Medical Biochemistry, Diagnostics, Bioinformatics and Clinical Research and Regulatory Affairs.

INDEX

 Index

Q

QMEAN, 24, 25, 31, 32, 33, 34, 38, 46, 48, 50, 51, 52, 53, 70, 78, 79, 80, 81, 82
QMEAN servers, 46
quantum dots, 126, 140

R

Ramachandran plot, 22, 24, 26, 27, 46, 48, 52, 67, 68, 70, 71, 72, 73, 74, 81
RAMPAGE, 22, 24, 68, 70
reactive oxygen species (ROS), 58

S

serum creatinine (SC), 11, 13, 15, 16, 17, 18, 20, 40, 53, 82, 83, 98, 108, 122, 142
SOD, 57, 59, 60, 61, 62, 146, 147, 149, 150, 151
solid models, 35, 36
SPSS, 14, 58, 60, 118
SPSS 12.0, 14, 58, 60
sulphonylureas, 6
superoxide, 59, 62, 65, 66, 97, 146, 147, 149
superoxide dismutase, 65, 66, 146, 147

T

T1DM, 3
T-cells, 115, 141
TCF7L2, v, 45, 46, 47, 48, 49, 50, 51, 52, 53, 54
TNF- α, 88, 91, 93, 94, 95, 100, 105, 106, 107, 116
total cholesterol (TC), 11, 13, 14, 15, 16, 17, 18, 99
toxicology, 3, 8

triglycerides, 14, 17, 88, 91, 93, 94, 95, 99, 100, 102, 103, 104, 105, 106
t-test, 14, 58, 60
tumor necrosis factor, 91, 99, 102
t-value, 93, 94, 105
type-1 diabetes, 2, 3, 6
type-2 diabetes (T2DM), 2, 3, 4, 22, 69, 89, 101, 114, 116, 122

U

UCLA-DOE, 24, 47, 70
uric acid, 111, 112, 113, 114, 115, 117, 118, 119, 120, 121, 122, 123

V

Verify3D, 24, 70
vitamin, v, vi, 11, 12, 13, 14, 15, 16, 17, 18, 19, 20, 57, 58, 59, 60, 61, 62, 63, 64, 65, 66, 107
vitamin C, v, 11, 12, 13, 14, 15, 16, 17, 18, 19, 20, 57, 58, 59, 60, 61, 62, 63, 64, 65, 107
vitamin E, vi, 12, 13, 14, 15, 16, 17, 19, 57, 58, 59, 60, 61, 62, 63, 64, 65, 66
vitamin E and C, 58, 59, 60, 63

W

WHATIF, 46, 48, 52

X

X-ray crystallography, 30, 70, 76, 82

Y

YASARA, 47, 48, 49

Lipid Rafts: Properties and Role in Signaling

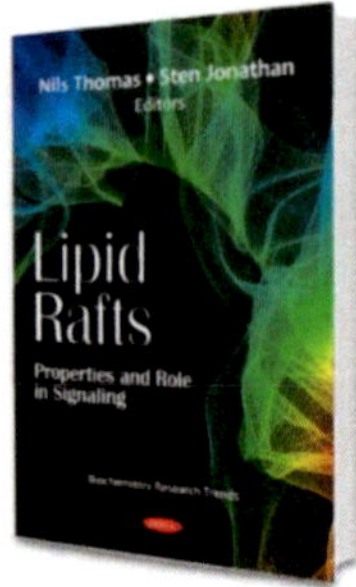

EDITORS: Nils Thomas and Sten Jonathan

SERIES: Biochemistry Research Trends

BOOK DESCRIPTION: Lipid rafts are nanometer-sized subdomains of the plasma membrane containing higher concentrations of cholesterol, phosphatidylinositols, and sphingolipids.

SOFTCOVER ISBN: 978-1-53613-624-1
RETAIL PRICE: $95

Trypsin: Anatomy, Biological Properties and Applications

EDITORS: Diana Faas and Joshua Holder

SERIES: Biochemistry Research Trends

BOOK DESCRIPTION: Trypsin, the protease with well-defined specificity, offers a great potential as a biocatalyst in numerous biomedical and industrial applications. In this collection, the authors discuss preparation and performance of trypsin immobilized on polysaccharide-based carriers.

SOFTCOVER ISBN: 978-1-53613-670-8
RETAIL PRICE: $82